Numerical Methods for Differential Systems

Recent Developments in Algorithms, Software, and Applications

ACADEMIC PRESS RAPID MANUSCRIPT REPRODUCTION

Numerical Methods for Differential Systems

Recent Developments in Algorithms, Software, and Applications

L. LAPIDUS
Princeton University

W. E. SCHIESSER
Lehigh University
and
Naval Air Development Center

ACADEMIC PRESS INC. New York San Francisco London 1976

A Subsidiary of Harcourt Brace Jovanovich, Publishers

ACADEMIC PRESS, INC.
111 Fifth Avenue, New York, New York 10003

United Kingdom Edition published by
ACADEMIC PRESS, INC. (LONDON) LTD.
24/28 Oval Road, London NW1

LIBRARY OF CONGRESS CATALOG CARD NUMBER: 76-22012

ISBN 0-12-4366406

PRINTED IN THE UNITED STATES OF AMERICA

Contents

List of Contributors

J. R. BARNEY, School of Engineering, Lakehead University, Thunderbay, Ontario

R. LEONARD BROWN, Department of Applied Mathematics and Computer Science, University of Virginia, Charlottesville, Virginia 22901

G. D. BYRNE, Department of Mathematics and Statistics and Department of Chemical and Petroleum Engineering, University of Pittsburgh, Pittsburgh, Pennsylvania 15260

M. B. CARVER, Atomic Energy of Canada, Limited, Chalk River Nuclear Laboratories, Chalk River, Ontario KOJ 1JO

F. DESCHARD, Department of Chemical Engineering, University of Connecticut, Storrs, Connecticut 06268

R. P. DICKINSON, JR., Lawrence Livermore Laboratory, P.O. Box 808, L-310, Livermore, California 94550

DOMINIC G. B. EDELEN, Center for the Application of Mathematics, Lehigh University, Bethlehem, Pennsylvania 18015

LENNART EDSBERG, Royal Institute of Technology, Department for Computer Sciences, Numerical Analysis, S-100 44 Stockholm 70, Sweden

W. H. ENRIGHT, Department of Computer Science, University of Toronto, Toronto, Ontario M5S 1A7

C. W. GEAR, Department of Computer Science, University of Illinois, Urbana, Illinois 61801

R. J. GELINAS, Science Applications, Inc., P.O. Box 34, 80 Mission Drive, Pleasanton, California 94566

DAVID R. HILL, Department of Mathematics, Temple University, Philadelphia, Pennsylvania 19121

A. C. HINDMARSH, Numerical Mathematics Group L-310, Lawrence Livermore Laboratory, P.O. Box 808, Livermore, California 94550

T. E. HULL, Department of Computer Science, University of Toronto, Toronto, Ontario M5S 1A7

A. I. JOHNSON, Faculty of Engineering Science, University of Western Ontario, London, Ontario

T. KOUP, Department of Chemical Engineering, University of Connecticut, Storrs, Connecticut 06268

F. T. KROGH, Mail Stop 125-109A, Jet Propulsion Laboratory, 4800 Oak Grove Drive, Pasadena, California 91103

W. LINIGER, IBM Thomas J. Watson Research Center, Yorktown Heights, New York 10598

NIEL K. MADSEN, Numerical Mathematics Group, Lawrence Livermore Laboratory, P.O. Box 808, L-310, Livermore, California 94550

R. MORGAN, Department of Chemical Engineering, University of Connecticut, Storrs, Connecticut 06268

M. R. SCOTT, Applied Mathematics Division 2642, Sandia Laboratories, Albuquerque, New Mexico 87115

L. F. SHAMPINE, Numerical Mathematics Division 5122, Sandia Laboratories, Albuquerque, New Mexico 87115

RICHARD F. SINCOVEC, Department of Computer Science, Kansas State University, Manhattan, Kansas 66506

L. F. STUTZMAN, Department of Chemical Engineering, University of Connecticut, Storrs, Connecticut 06268

H. A. WATTS, Applied Mathematics Division 2642, Sandia Laboratories, Albuquerque, New Mexico 87115

Preface

The numerical solution of large-scale scientific and engineering problems, expressed as systems of ordinary and partial differential equations (ODEs and PDEs, respectively), is now well established. The insight provided by this type of analysis is considered indispensable in the evaluation and design of advanced technology systems. Thus, methods for improving and extending the application of numerical computation to the solution of ODE/PDE systems is an active area of research. The papers in this volume cover a spectrum of recent developments in numerical algorithms for ODE/PDE systems: theoretical approaches to the solution of nonlinear algebraic and boundary value problems via associated differential systems, new integration algorithms for initial-value ordinary differential equations with particular emphasis on stiff systems (i.e., systems with widely separated eigenvalues), finite difference algorithms particularly suited for the numerical integration of PDE systems, general-purpose and special-purpose computer codes for ODE/PDEs, which can be used by scientists and engineers who wish to avoid the details of numerical analysis and computer programming, and user experience both with these particular developments and generally within the field of numerical integration of differential systems as reported by a panel of recognized researchers.

The papers in this volume were first presented in a four-part symposium at the 80th National Meeting of the American Institute of Chemical Engineers (A.I.Ch.E.), in Boston, September 7–10, 1975. Although some of the papers are oriented toward applications in chemistry and chemical engineering, most generally relate to new developments in the computer solution of ODE/PDE systems.

The papers by Liniger, Hill, and Brown present new algorithms for initial-value, stiff ODEs. Liniger's algorithms are A-stable and achieve accuracy up to sixth order by averaging A-stable second-order solutions. Thus the method is well suited for the parallel integration of stiff systems.

Hill's second derivative multistep formulas are based on g-splines rather than the usual polynomial interpolants. Brown's variable order, variable stepsize algorithm is A-stable for orders up to seven, but requires the second and third derivatives of the solution; it is presented essentially for linear systems, but extensions to nonlinear systems are discussed.

Recent research in stiff systems has produced a large number of proposed numerical algorithms; some newer algorithms have already been mentioned. Thus the field has developed to the point that comparative evaluation is necessary to determine which contributions are most useful for a broad spectrum of problem systems. Enright and Hull have tested a selected set of recently reported algorithms on a collection of ODEs arising in chemistry and chemical engineering. They give recommendations based on the results of these tests to assist the user in selecting an algorithm for a particular stiff ODE problem system.

The two papers by Edelen discuss the interesting concept that a differential system can be integrated to an equilibrium condition to obtain a solution to a problem system of interest. For example, a nonlinear algebraic or transcendental system has a special-case solution of a related initial-value ODE system. Similarly, boundary-value problems can be solved by integrating associated initial-value problems to equilibrium. Methods for constructing the related initial-value problem are presented which have limit solutions for the system of interest. The convergence may be in finite time as well as the usual large-time exponential convergence.

Even though the mathematical details of new, efficient algorithms for stiff differential systems are available, their practical implementation in a computer code must be achieved before a user community will readily accept these new methods. Codes are required that are user-oriented (i.e., can be executed without a detailed knowledge of the underlying numerical methods and computer programming), thoroughly tested (to give reasonable assurance of their correctness and reliability), and carefully documented (to give the user the necessary information for their use). Several general-purpose codes for stiff ODE systems have been developed to meet these requirements.

The DYNSYS 2.0 system by Barney and Johnson, and the IMP system by Stutzman *et al.* include translators that accept problem-oriented statements for systems modeled by initial-value ODEs and then perform the numerical integration of the ODEs by implicit algorithms to achieve computational efficiency for stiff systems.

Hindmarsh and Byrne describe a FORTRAN-IV system, EPISODE, which is also designed to handle stiff systems. EPISODE can be readily incorporated into any FORTRAN-based simulation and does not require translation of input code provided by the user.

Application of all three systems to problems in chemistry and chemical engineering are presented. A particular application of the EPISODE system to atmospheric kinetics is described by Dickinson and Gelinas. Their system consists of two sections: a code for generating a system of initial-value ODEs and its Jacobian matrix from user-specified sets of chemical reaction processes and the code for numerical integration of the ODEs.

Edsberg describes a package especially designed for stiff problems in chemical kinetics, including a parameter estimation feature. The design of the system is based on the specific structure of chemical reaction system equations obeying mass action laws.

All the preceding systems are for initial-value ODEs. Scott and Watts describe a system of FORTRAN-based, transportable routines for boundary-value ODEs. These routines employ an orthonormalization technique, invariant imbedding, finite differences, collocation, and shooting.

Finally in the area of PDEs, recent emphasis has been on the application of the numerical method of lines (NMOL). Basically, a system of PDEs containing partial derivatives with respect to both initial-value and boundary-value independent variables is replaced by an approximating set of initial-value ODEs. This is accomplished by discretizing the boundary-value or spatial partial derivatives. The resulting system of ODEs is then numerically integrated by an existing initial-value stiff systems algorithm. An important consideration in using the NMOL is the approximation of the spatial derivatives. Madsen and Sincovec relate some of their experiences with this problem in terms of a general-purpose FORTRAN-IV code for the NMOL. Also, Carver discusses an approach for the integration of the approximating ODEs through a combination of a stiff systems integrator and sparse matrix techniques. Basic considerations in the descretization of the spatial derivatives are also considered by Carver.

The volume concludes with the comments from a panel of experts chaired by Byrne. These statements reflect extensive experience in the solution of large-scale problems and provide an opportunity for the reader to benefit from this experience.

Most of the contributions in this volume are relevant to the solution of large-scale scientific and engineering problems in general. Thus these new developments should be of interest to scientists and engineers working in a spectrum of application areas. In particular, several of the codes are available at nominal cost or free of charge, and they have been written to facilitate transportability. The reader can readily take advantage of the substantial investment of effort made in the development, testing, and documentation of these codes. Details concerning their availability can be obtained from the authors.

L. Lapidus
W. E. Schiesser

High-Order A-Stable Averaging Algorithms for Stiff Differential Systems

W. Liniger

1. In this paper various integration methods for systems of stiff differential equations,

$$\dot{y} = f(y), \tag{1}$$

are proposed which are A-stable and of order of accuracy p up to six. Since for linear multistep formulas (LMF),

$$\sum_{j=0}^{k} \alpha_j x_{n+j} - h \sum_{j=0}^{k} \beta_j f(x_{n+j}) = 0, \tag{2}$$

A-stability is compatible only with orders of accuracy not exceeding two [1], the present methods cannot be of LMF type. Instead they consist of 1) integrating a given problem several times by an A-stable LMF (thus of order $p \leq 2$) which contains free parameters (other than the integration step h), each solution corresponding to a different set of parameter values; and 2) forming a suitable linear combination, loosely referred to

---

† Research sponsored in part by the Air Force Office of Scientific Research (AFOSC), United States Air Force, under Contract No. F44620-75-C-0058. The United States Government is authorized to reproduce and distribute reprints for governmental purposes notwithstanding any copyright notation hereon.

as an "average"*, of the different LMF solutions. The idea of averaging was first proposed in [2] and is related to Richardson extrapolation [1] with respect to h. In writing program packages for the solution of stiff systems of differential equations, A-stable methods such as the present ones might be substituted for methods which are only stiffly stable such as the backward differentiation formulas. The latter, while successful in many problems, have in fact been found to lead to failures in some cases because of their lack of A-stability.

The algorithms proposed in [2] are restricted to LMF of Adams type ($\alpha_j = 0$, $j=0,\ldots,k-2$) and to achieving an order of the averaged solution not exceeding $2p \leq 4$, where $p \leq 2$ is the order of accuracy of the underlying A-stable LMF. However, in [2] it is also stated that, and by an example, shown how the averaging method can be developed for arbitrary LMF and to arbitrary orders $> 2p$. This is accomplished by studying the dependence on the formula parameters of the asymptotic expansion of the global truncation error. A technique for computing that expansion has been developed by the author and will be reported elsewhere. The present set of methods is the result of applying that technique to the four-parameter class of A-stable LMF with step number k=3 and then specializing to a particularly interesting two-parameter sub-class thereof.

---

* It is not an average in the sense that some of the "weights" can be negative.

A LMF which is A-stable for <u>some</u> values of parameters on which it depends is usually not A-stable for <u>all</u> parameter values. Instead it possesses a domain of A-stability in parameter space from which the parameter sets of all of the formulas being averaged have to be selected in order for the averaged, high-order solution to be A-stable. The discussion of A-stability can be complicated if the formula is not suitably parametrized. In particular, if some of the coefficients α_j and β_j, or (as in [2]) quantities on which these coefficients depend linearly, are used as parameters, then the A-stability analysis is quite difficult. A better parametrization was proposed in [3] where an explicit construction is given, for all k, of a class of A-stable LMF of order p = 2 depending on the maximum number possible, 2k-2, of free parameters. In general, not all A-stable formulas with these specifications are found by this construction because it is based on conditions which are only sufficient for A-stability. However, by the considerations of [3] it is easy to verify that, for $k \leq 3$, those conditions are necessary as well. The four-parameter class mentioned in the previous paragraph is thus the class of <u>all</u> A-stable LMF of that type. In the new parameters of [3] the A-stability domain is simply the Cartesian product of the first octant in a three-dimensional subspace and of an interval with respect to the fourth parameter. The use of the parameters proposed in [3] greatly simplifies also the asymptotic analysis of the global truncation error mentioned above.

2. Denoting, respectively, by r,u,v, and w the parameters $\gamma_1, \gamma_2, \gamma_3$ and e_1 defined in [3], the four-parameter family of all LMF with p = 2, k = 3, can be written as

$$\begin{aligned}
&[2(-r+u-v) + \frac{w}{u}]\, x_n + [2(3r-u-v) - \frac{3w}{u}]\, x_{n+1} \\
&\quad + [2(-3r-u+v) + \frac{3w}{u}] x_{n+2} + [2(r+u+v - \frac{w}{u}] x_{n+3} \\
&\quad -h[(-1+r-u+v + \frac{v}{u})\dot{x}_n + (3-r-u+3v - \frac{v}{u})\dot{x}_{n+1} \qquad (3) \\
&\quad + (-3-r+u+3v - \frac{v}{u})\dot{x}_{n+2} + (1+r+u+v + \frac{v}{u})\dot{x}_{n+3}] = 0.
\end{aligned}$$

The LMF defined by (3) is A-stable if and* only if [3]

$$r>0, u>0, v>0, \qquad 0\leq w\leq 2ru. \qquad (4)$$

For a moment, we shall denote by y the value of an exact solution y(t) of the differential equation (1) at a generic gridpoint $t = t_n = nh$, n integer. Similarly, we denote by x the value x_n, associated with t_n, of an approximate solution to (1) computed with appropriate initial conditions.** Finally let $\varepsilon = x-y$ denote the global truncation error. It can then be shown that, for sufficiently differentiable f's and for any LMF of order p, the global truncation error has an asymptotic expansion of the form

$$\begin{aligned}
\varepsilon = &\sum_{i=0}^{p-1} h^{p+i} T_i Y_i + \sum_{i=0}^{p-1} h^{2p+i} [T_{p+i} Y_{p+i} \\
&+ \sum_{j=0}^{i} T_j T_{i-j} Y_{j,i-j}] + O(h^{3p}), \qquad (5)
\end{aligned}$$

---

* Except possibly for "borderline" cases where the polynomial $\sigma(\zeta) = \sum_{j=0}^{3} \beta_j \zeta^j$ has roots on the unit circle, which is equivalent to saying that one or more of the quantities r,u,v may be equal to zero.

** As described in section 4 below.

where the quantities Y_i and Y_{ij} depend only on the exact solution y but not on the integration formula and the T_i are functions of formula parameters. For the particular LMF defined by (3), the expansion (5) specializes to

$$\varepsilon = h^2 T_0 Y_0 + h^3 T_1 Y_1 + h^4 (T_2 Y_2 + T_0^2 Y_{0,0}) + h^5 [T_3 Y_3 + T_0 T_1 (Y_{0,1} + Y_{1,0})] + O(h^6), \tag{6}$$

where

$$T_0 = \frac{1}{u} + \frac{1}{3} + \frac{w}{2uv}, \tag{7}$$

$$T_1 = \frac{w}{2v^2}, \tag{8}$$

$$T_2 = \frac{1}{3} T_0 + \left(\frac{r}{uv} - \frac{4}{45}\right) + w\left(-\frac{u}{2v^3} + \frac{r}{2uv^2} - \frac{w}{4u^2v^2} - \frac{1}{2u^2v}\right), \tag{9}$$

$$T_3 = \frac{r}{v^2} - w\left(\frac{1}{2uv^2} - \frac{1}{3v^2} - \frac{r}{v^3} + \frac{u^2}{2v^4} + \frac{w}{2uv^3}\right). \tag{10}$$

One confirms immediately that $T_0 \neq 0$ if the A-stability conditions (4) are enforced and the order of the LMF (3) thus cannot exceed two*.

The idea of averaging can now be formalized as follows. Consider m individual formulas of the class (3), each corresponding to a different set of parameters $S_j = (r_j, u_j, v_j, w_j)$, $j=1,2,\ldots,m$. Denote by x^j, ε^j, and T_i^j, $i=0,1,2,3$, the approximate solution x, the error ε, and the auxiliary quantities T_i, respectively, corresponding to S_j. Now form the linear combination

$$z = \sum_{j=1}^{m} \nu_j \, x^j, \tag{11}$$

---

* We should be reminded that, for an LMF of order p, the local truncation error is $O(h^{p+1})$ whereas, over a finite interval of integration, the global error is $O(h^p)$.

where the "weights" ν_j satisfy the first m of the relations

$$\sum_{j=1}^{m} \nu_j = 1, \tag{12}$$

$$\sum_{j=1}^{m} \nu_j T_0^j = 0, \tag{13}$$

$$\sum_{j=1}^{m} \nu_j T_1^j = 0, \tag{14}$$

$$\sum_{j=1}^{m} \nu_j T_2^j = 0, \tag{15}$$

$$\sum_{j=1}^{m} \nu_j (T_0^j)^2 = 0, \tag{16}$$

$$\sum_{j=0}^{m} \nu_j T_3^j = 0, \tag{17}$$

$$\sum_{j=0}^{m} \nu_j T_0^j \; T_1^j = 0. \tag{18}$$

Then, because of condition (12), quantities independent of the change of parameters, such as y, are left unchanged in forming the average. Therefore, $z = \sum_{j=1}^{m} \nu_j x^j = y + \sum_{j=1}^{m} \nu_j \varepsilon^j$ and, for m=2,3,5, and 7, we have $z-y = O(h^{\mu})$, μ=3,4,5, and 6, respectively, so that z represents a solution of order μ. Furthermore, if all the S_j are picked from the A-stability domain (4), then all x^j as well as z are A-stable.

It may now be observed by considering the relations (7) through (10) that the averaging procedure becomes greatly simplified for a two-parameter subclass of the class of formulas defined by (3). In fact if we let

$$w = 0, \tag{19}$$

then $T_1 = 0$ and averaging $m = 2$ solutions will produce a solution z of order four. If in addition to (19) we let

$$r = 4uv/45, \tag{20}$$

then T_0 and T_2 are proportional and are averaged out together. Thus, in this case, if $m = 3$ and if the conditions (12), (13), and (16) are imposed, then z is of order five. Finally, again subject to (19) and (20), averaging $m = 4$ solutions with the weights ν_j satisfying (12), (13), (16), and (17) will result in a sixth-order solution z as (19) implies (18). If we substitute for the T_i the expressions given by (7) through (10), subject to the constraints (19) and (20), then the reduced set of averaging constraints in terms of the remaining parameters u and v can be written as

$$\sum_{j=1}^{m} \nu_j = 1, \tag{12}$$

$$\sum_{j=1}^{m} \left(\frac{1}{u_j}\right)\nu_j = -\frac{1}{3}, \tag{13'}$$

$$\sum_{j=1}^{m} \left(\frac{1}{u_j}\right)^2\nu_j = \frac{1}{9}, \tag{16'}$$

$$\sum_{j=1}^{m} \left(\frac{u_j}{v_j}\right)\nu_j = 0. \tag{17'}$$

The reduced formula (3) becomes

$$\begin{aligned}&(2u-2v - \frac{8uv}{45})x_n + (-2u-2v + \frac{24uv}{45})x_{n+1}\\ &\quad + (-2u+2v - \frac{24uv}{45})x_{n+2} + (2u+2v + \frac{8uv}{45})x_{n+3}\\ &-h[(-1-u+v + \frac{v}{u} + \frac{4uv}{45})\dot{x}_n + (3-u+3v - \frac{v}{u} - \frac{4uv}{45})\dot{x}_{n+1}\\ &\quad + (-3+u+3v - \frac{v}{u} - \frac{4uv}{45})\dot{x}_{n+2} + (1+u+v + \frac{v}{u} + \frac{4uv}{45})\dot{x}_{n+3}] = 0.\end{aligned} \tag{21}$$

One might ask whether the sets S_j could be chosen so that some of the relations (12), (13'), (16'), and (17') are satisfied simultaneously and thus the m needed to achieve certain orders could be further reduced. This is indeed possible but may be incompatible with the A-stability constraints (4). For example, for $m = 2$, u_1 and u_2 can be so chosen that (13') and (16') are satisfied simultaneously but then u_1 and u_2 cannot both be non-negative.

3. Thus far the choice of the parameters u and v is subject only to the A-stability conditions (4). But, whereas the qualitative statements made above about the order of accuracy do not depend on u and v, quantitative measures for the error ε do, of course, depend on the values of these parameters. One might, for any given set of differential equations, try to choose a suitable pair u,v say, by exponential fitting [4,5]. Or one might try to optimize this choice a priori once and for all so as to minimize, globally with respect to h, some measure of the local

truncation error, as was done in [6] for the weighted Euler formula using an L_∞-norm. An approach similar to the one taken in [6] but using the L_2-norm predicts that pairs of parameter values $(u,v(u))$, where

$$v(u) = (2.6432u + 2.7699u^2 + 0.8566u^3 + 0.0581u^4)/ \qquad (22)$$
$$2(1.5222 + 1.5535u + 0.5247u^2 + 0.0620u^3 + 0.0024u^4),$$

are likely to produce relatively small truncation errors for the formula (21). It should be remarked, however, that averaging solutions of (21), each of which has associated with it a small local truncation error, is not necessarily best for reducing the error in the averaged solution z. The reason is that that error depends on the formula parameters also via the weights ν_j. This relationship is a complicated one and as yet not well understood. The parameter values used in the test calculations described below were chosen partly based on (22), partly found by numerical experimentation.

4. The performance of the averaging procedure defined above crucially depends on the choice of the starting values for the underlying difference solutions $\{x_n^j\}$. We give a method for finding suitable starting values which, for simplicity, we explain for a two-by-two system of differential equations. To justify this method, we consider a particular solution y(t) of (1) and associate with it the differential equation

$$\dot{y} = f(y(t)) + J(y(t))\,[y-y(t)]\ , \ J = \partial f/\partial y, \qquad (23)$$

which is a linear approximation to (1) in the neighborhood of $y = y(t)$ in the sense that the right side of (23) represents, up to and including the linear term, the Taylor-expansion of $f(y)$ with respect to y around $y = y(t)$. In the neighborhood of $t = 0$ we further approximate (23) by

$$\dot{y} = Jy + c, \tag{24}$$

where $J = J(y_0)$, $y_0 = y(0)$ is the initial condition of (1), and

$$c = f(y_0) - J(y_0)y_0. \tag{25}$$

Now we apply the formula (21) to (24). Written in terms of the well-known polynomials $\rho(\zeta) = \sum_{j=0}^{3} \alpha_j \zeta^j$ and $\sigma(\zeta) = \sum_{j=0}^{3} \beta_j \zeta^j$ and of the shift operator E defined by $Ex(t) = x(t+h)$, the difference equation becomes

$$[I\rho(E) - hJ\sigma(E)]x_n = h\sigma(1)c. \tag{26}$$

It possesses a constant particular solution, $x_n \equiv \xi$, defined by $[\rho(1)I - h\sigma(1)J]\xi = h\sigma(1)c$. By the consistency and the relative primality of $\rho(\zeta)$ and $\sigma(\zeta)$, both of which we assume, we have $\rho(1) = 0$ and $\sigma(1) \neq 0$. Thus,

$$\xi = - J^{-1}c. \tag{27}$$

Then, if we write $x_n = \xi + q_n$, q_n satisfies

$$[I\rho(E) - hJ\sigma(E)]q_n = 0. \tag{28}$$

Now assume that J has a "small" eigenvalue $\lambda_1 < 0$ (in the sense that $|h\lambda_1| << 1$) and a "large" eigenvalue $\lambda_2 < 0$ (for which $|h\lambda_2| >> 1$). These eigenvalues are associated with the "smooth" and the stiff components of the differential solution, respectively. For $i=1,2$, let ζ_{ij}, $j=1,\ldots,3$, be the roots of the characteristic polynomial

$\rho(\zeta) - h\lambda_i\sigma(\zeta)$. Then, provided the ζ_{ij} are all distinct, the general solution of (28) takes the form

$$q_n = \sum_{i=1}^{2} \left(\sum_{j=1}^{3} \gamma_{ij}\zeta_{ij}^n \right)\eta_i, \quad n=0,1,\ldots, \tag{29}$$

where η_1 and η_2 are linearly independent unit-eigenvectors of J associated with the real distinct eigenvalues λ_1,λ_2, respectively. Let, in particular, $\zeta_{1,1}$ be the principal root which, by virtue of the second order accuracy of the formula (21), approximates $e^{\lambda_1 h}$ to $O(h^2)$ in the limit $h \to 0$.

The difference solution (27) is also a solution of the differential equation (24). The complete solution of (24) can be written in the form $y(t) = \xi + s(t)$, where

$$s(t) = \sum_{i=1}^{2} \delta_i e^{\lambda_i t}\eta_i. \tag{30}$$

We say that s(t) is smooth if $\delta_2 = 0$; i.e., if s(t) has no component in the direction of the eigenvector η_2 of the rapidly changing solution. In this case,

$$s(t) = \delta_1 e^{\lambda_1 t}\eta_1. \tag{31}$$

Since $y_0 - \xi = s(0)$, this means that $y_0 - \xi$ is proportional to η_1. By (25) and (27), $y_0 - \xi = y_0 + J^{-1}[f(y_0) - Jy_0] = J^{-1}f(y_0)$. Thus, $f(y_0)$ is in turn proportional to η_1. Therefore, in order for y(t) to be smooth, the initial vector y_0 must satisfy

$$J(y_0)f(y_0) = \lambda_1 f(y_0), \tag{32}$$

meaning geometrically that the tangent to the solution curve is parallel to η_1.

The asymptotic error expansion (5) is valid only in the limit $h\lambda \to 0$, where λ is any given eigenvalue of J, and only for solutions $\{x_n\}$ of the difference equation (2) which, locally near t = 0, are associated with the principal root of the characteristic polynomial $\rho(\zeta)-h\lambda\sigma(\zeta)$. Since $|h\lambda_2|>>1$, the presence of the solutions $\{\zeta_{2,j}^n\}$ in the representation for q_n (equation (29)) must degrade or invalidate that asymptotic expansion and the averaging method based on it, unless the coefficients $\gamma_{2,j}$ are small compared to the absolute accuracy level which one attempts to achieve. The same is true for the parasitic (extraneous) solutions $\{\zeta_{1,j}^n\}$, j=2,3. It is therefore natural to choose the initial data for the difference solution $\{x_n\}$ in such a way as to make $\gamma_{1,2}= \gamma_{1,3}= \gamma_{2,1}= \gamma_{2,2}= \gamma_{2.3}= 0$. In this case,

$$q_n = \gamma_{1,1}\zeta_{1,1}^n\eta_1. \tag{33}$$

Now if one makes the natural choice

$$x_0 = y_0, \tag{34}$$

then $(x_0-\xi) - (y_0-\xi) = q_0- s(0) = 0$, i.e., $q_0 = s(0)$; this is consistent with relations (31) and (33) which state that both q_0 and s(0) are in the direction of η_1. In conclusion, suitable initial data for the difference solution are defined by

$$x_n = \xi + \zeta_{1,1}^n(y_0-\xi), \qquad n= -2,-1,0, \tag{35}$$

where y_0 is an initial vector for the differential equation satisfying the smoothness condition (32), ξ is defined by (27) via (25), and $\zeta_{1,1}$ is the principal root of $\rho(\zeta)-h\lambda_1\sigma(\zeta)$, where λ_1 is the "small" root of the Jacobian matrix $J = J(y_0)$.

5. The algorithm, defined by formula (21) and the averaging relations (12), (13'), (16'), and (17'), was implemented in a way similar to that described in [2]. It was then applied to the two test problems referred to as TP1 and TP2. The problem TP1:

$$\begin{aligned} \dot{y}_1 &= -2000y_1 + 1000y_2 + 1000, & y_1(0) &= 0, \\ \dot{y}_2 &= y_1 - y_2, & y_2(0) &= 0, \end{aligned} \tag{36}$$

whose exact solution is

$$y(t) = \frac{1}{\lambda_1-\lambda_2}\begin{pmatrix} \lambda_2(1+\lambda_1) & -\lambda_1(1+\lambda_2) \\ \lambda_2 & -\lambda_1 \end{pmatrix}\begin{pmatrix} e^{\lambda_1 t} \\ e^{\lambda_2 t} \end{pmatrix} + \begin{pmatrix} 1 \\ 1 \end{pmatrix}, \tag{37}$$

is identical with problem P1 of [2] and was proposed originally in [7]. The (constant) eigenvalues are

$\lambda_{1,2} = \frac{1}{2}[-2001 \pm (2001^2 - 4000)^{1/2}]$, $\lambda_1 \approx -0.499875$,

$\lambda_2 \not\approx -2000.500125$ with a "stiffness ratio" of ≈ 4000. The initial condition at $t = 0$ in this case does not define a smooth solution according to the criterion (32). Consequently, the problem TP1 was solved numerically over the interval $1 \le t \le 4$ which lies in the asymptotic phase of the solution (37) during which the stiff component varying like $e^{\lambda_2 t}$ is negligibly small. The starting vector $x_0 = y_0 = y(1)$, whose components are

$x_{0,1} \approx 0.696545108$, $x_{0,2} \approx 0.3932419055$, was computed from (37); it satisfies the smoothness criterion (32) to within exponentially small corrections.

The nonlinear problem TP2 is defined by

$$\begin{aligned} \dot{y}_1 &= -\tfrac{1}{5}[(4b+g)y_1 + (2b-2g)y_2] - \tfrac{2}{25}\mu e^{bt}(2y_1+y_2)^2, \\ \dot{y}_2 &= -\tfrac{1}{5}[(2b-2g)y_1 + (b+4g)y_2] - \tfrac{\mu}{25} e^{bt}(2y_1+y_2)^2. \end{aligned} \qquad (38)$$

With the initial conditions

$$\begin{aligned} y_1(0) &= 2 - a, \\ y_2(0) &= 1 + 2a, \end{aligned} \qquad (39)$$

the equations (38) have the exact solution

$$\begin{aligned} y_1(t) &= 2F(t) - ae^{-gt} \\ y_2(t) &= F(t) + 2ae^{-gt}, \end{aligned} \qquad (40)$$

where

$$F(t) = \tfrac{1}{5}[2y_1(t) + y_2(t)] = e^{-bt}/(1+\mu t). \qquad (41)$$

The eigenvalues of the Jacobian matrix of equations (38), evaluated on the solution (40), are

$$\begin{aligned} \lambda_1 &= -[b+2\mu e^{bt}F(t)], \\ \lambda_2 &= -g. \end{aligned} \qquad (42)$$

Thus, for positive values of b, μ, and g, the solution is asymptotically stable for all $t \geq 0$. Furthermore, if $g \gg b$ and μ, TP2 is stiff. In the calculations, $b = 0.2$ and $g = 200$ were used, giving a stiffness ratio of ≈ 1000. In the limit $\mu = 0$, TP2 is linear with constant coefficients. By increasing μ from $\mu = 0$, the effect of the nonlinearity may be studied. In the calculations, an approximation was sought to the particular solution (40) which is defined by the initial conditions $y_1(0) = 2$ and $y_2(0) = 1$. This particular solution is "smooth"; i.e., the stiff term, ae^{-gt}, vanishes because its coefficient is $a = 0$ (any other solution has a boundary layer of amplitude a at $t = 0$). In fact, the starting vector $x_0 = \binom{2}{1}$ rigorously satisfies the smoothness condition (32). For the test problem TP2, a smooth starting vector is arrived at by inspection. In general, it must be found by solving the nonlinear equations (32), say by Newton's method with an initial guess which might be gotten from a smooth numerical solution which has been carried into the asymptotic phase by integrating through the boundary layer with small enough h.

The numerical results of applying the present averaging algorithm to TP1, and to TP2 with $\mu = 0$, $\mu = 10^{-5}$, and $\mu = 10^{-4}$, are plotted in figures 1 through 4, respectively. In each figure, the numbering of the curves corresponds to the number m of solutions averaged. The underlying solutions $\{x_n^j\}$ were computed for the pairs of parameter values (u_j, v_j), $j=1,\ldots,m$, $m \leq 4$, where $u_1 = 1$, $u_2 = 1.5$, $u_3 = 10$, and $u_4 = 30$ and where $v_j = v(u_j)$, $j=1,\ldots,4$ with $v(u)$ defined by (22). Specifically, $v_1 = 0.8633$, $v_2 = 1.2733$, $v_3 = 5.5819$, and $v_4 = 8.7238$. For TP2, the intervals of integration were $0 \leq t \leq 5$ for $\mu = 0$ and 10^{-5}, and $0 \leq t \leq 12$ for $\mu = 10^{-4}$. The values plotted represent mean square relative errors over an output set of solution values associated with certai subsets of the set of gridpoints. For TP1, the error in the first component of the solution vector is plotted; for TP2, the relative error happens to be the same for both components. The slopes of th error curves No. 2, 3 and 4 are expected to be equal to the orders 4, 5, and 6 of the global error of the averaged solution z which the theory predicts for m = 2, 3, and 4, respectively. The curves No. 1 represent the global errors in the unaveraged solution $\{x_n^1\}$ whose order is two; that order, too, should be reflected in the slopes of those curves.

The results plotted in figures 1 and 2, which correspond to linear problems with constant coefficients, clearly confirm the averaging theory as far as orders are concerned. They also show that, for given, not too large values of h, the errors decrease absolutely as the number m of averaged solutions increases.

For linear problems, such as TP1, and TP2 for $\mu = 0$, the method described above for selecting smooth starting data for the difference equation remains effective over the whole interval of integration. For nonlinear problems, however, the different solution modes seem to interact and the non-smooth solutions seem to creep back in after a while. Therefore, with the present implementation of the algorithm, averaging may be expected to work only above a certain absolute error level; below that

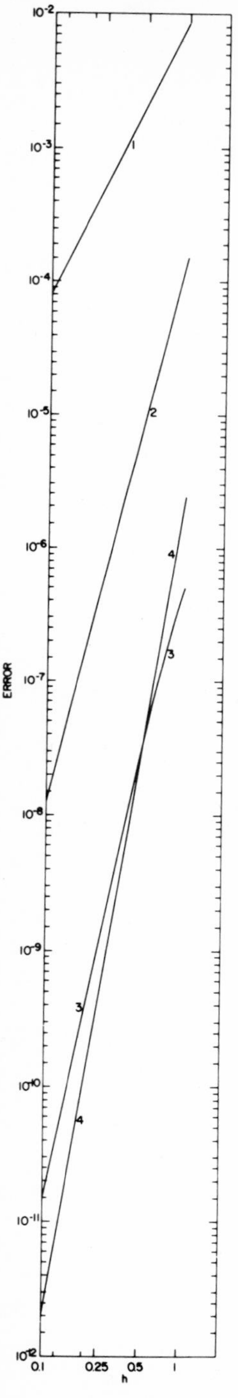

Fig. 1

Relative errors for problem TP1.

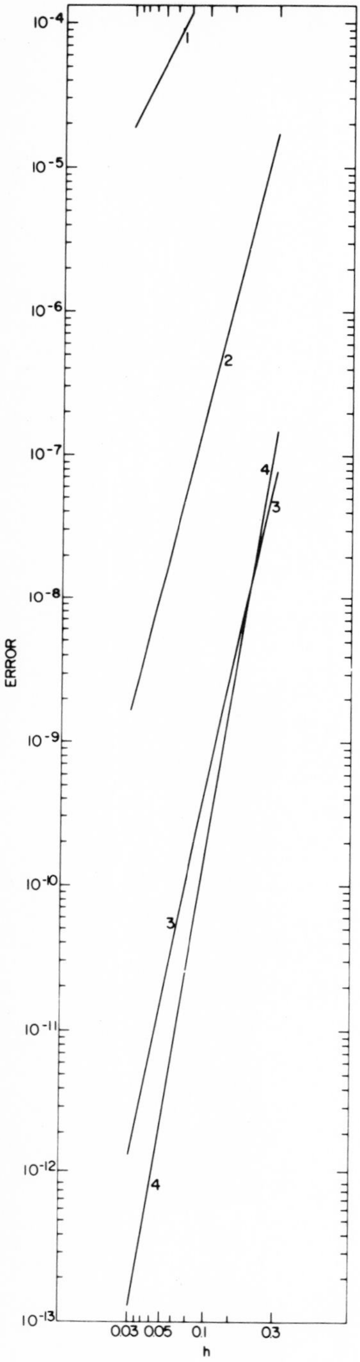

Fig. 2

Relative errors for problem TP2 with $\mu=0$.

level the overall error is dominated by the presence of the non-smooth solutions. With the present starting procedure, these solutions come in locally with an amplitude which is $O(h^3)$, the order of the local truncation error of the underlying solutions, and globally to $O(h^2)$. For the nonlinear versions of problem TP2, this situation is exhibited in figures 3 and 4 where, for not too small values of h and absolute error levels, the steepness of the error curves and their relative position correctly exhibit the effect of averaging, whereas below those levels the slopes are approximately 2 or 3 and, for given values of h, the absolute errors cannot be lowered below certain thresholds by increasing m. If a problem is weakly nonlinear, as is TP2 for the non-zero values of μ used herein, then starting the numerical solution with "smooth" data seems to be enough to make the averaging procedure work within the limits mentioned above. For more strongly nonlinear problems, a suitable smoothing method may have to be applied from time to time. Conceivably, an averaging recipe could also be derived from a complete asymptotic expansion of the global error which accounts for the presence of the non-smooth solutions.

In conclusion we remark that, as in Richardson extrapolation [1], the solutions x^j, $j=1,...,m$, underlying the formation of

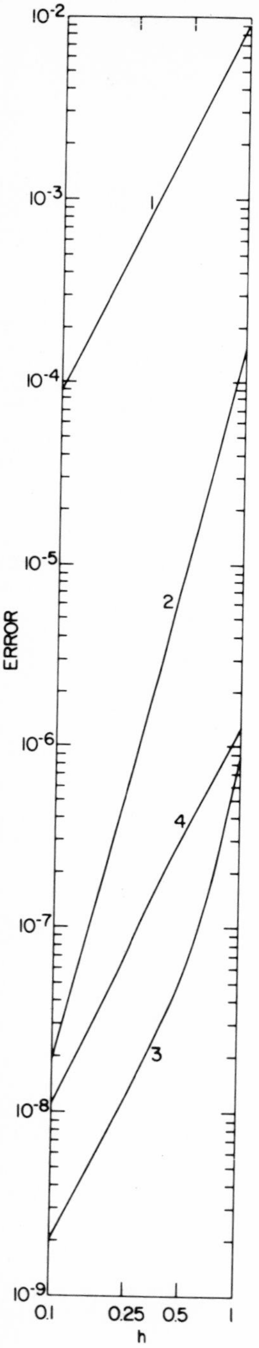

Fig. 3

Relative errors for problem TP2 with $\mu = 10^{-5}$.

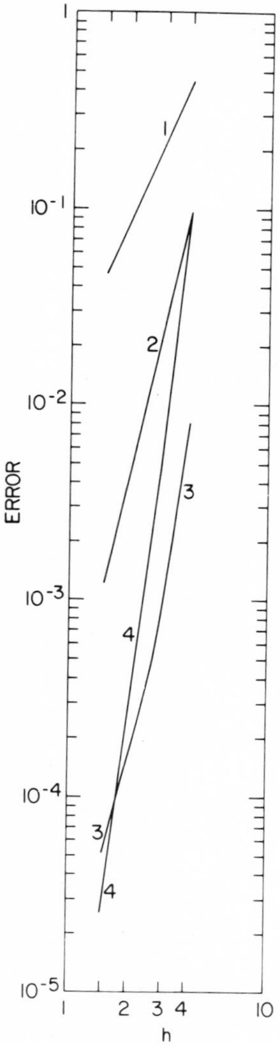

Fig. 4

Relative errors for problem TP2 with $\mu = 10^{-4}$.

the averaged solution z must be computed independently of one another and of z in order to preserve A-stability. Therefore, averaging algorithms lend themselves naturally to parallel processi

Acknowledgment: I would like to thank my colleague, Dr. F. Odeh, for the valuable discussions we had concerning this extension of ou earlier joint work.

REFERENCES

[1] G. G. Dahlquist, "A special stability criterion for linear multistep methods," BIT 3 (1963), 22-43.

[2] W. Liniger and F. Odeh, "A-stable, accurate averaging of multistep methods for stiff differential equations," IBM J. Res. Develop., 16 (1972), 335-348.

[3] W. Liniger and T. Gagnebin, "Construction of a family of second order, A-stable k-step formulas depending on the maximum number, 2k-2, of parameters," Stiff Differential Systems (R. A. Willoughby, Ed.), Plenum Press, New York, (1974), 217-227.

[4] W. Liniger and R. A. Willoughby, "Efficient integration methods for stiff systems of ordinary differential equations," IBM Research Report, RC-1970 (Dec. 20, 1967).

[5] E. F. Sarkany and W. Liniger, "Exponential fitting of matricial multistep methods for ordinary differential equations," Math. Comp. 28 (1974), 1035-1052.

[6] W. Liniger, "Global accuracy and A-stability of one- and two-step integration formulae for stiff ordinary differential equations," Lecture Notes in Mathematics 109, Springer Verlag, Berlin, (1969) 188-193.

[7] M. E. Fowler and R. M. Warten, "A numerical integration technique for ordinary differential equations with widely separated eigenvalues," IBM J. Res. Develop., 11 (1967), 537-543.

Second Derivative Multistep Formulas Based on g-Splines

David R. Hill

1. INTRODUCTION

This paper is concerned with the construction of second derivative multistep formulas for the numerical solution of stiff differential equations. Enright in [4] and [5] has developed several classes of such formulas and implemented them into numerical methods. Here we shall construct another class of such formulas, based on g-splines, and compare them with Enright's formulas in several standard ways. The motivation for such a construction comes from similar work for non-stiff ordinary differential equations in [1] and [10] and for functional differential equations in [8] and [9].

2. NOTATION AND DEFINITIONS

We consider the following class of k-step second derivative formulas

$$(1)\quad \sum_{i=0}^{k} \alpha_i y_{n+i} = h\{\sum_{i=0}^{k} \beta_i f_{n+i}\} + h^2\{\sum_{i=0}^{k} \gamma_i f'_{n+i}\} \quad n = 0,1,2,\ldots$$

for the numerical solution of

$$(2)\qquad y' = f(x,y),\ y(a) = \eta$$

where k is a fixed integer, $f_m \equiv f(x_m,y_m)$, $f'_m \equiv \frac{d}{dx} f(x_m,y_m)$ and $y_m \simeq y(x_m)$, the true solution of (2).

To define the order of a formula such as (1) we use a natural extension of the definition given by Henrici [7;p.221]. We shall say that second derivative multistep formula (1) is of order p if $E_0 = E_1 = \ldots = E_p = 0$, but $E_{p+1} \neq 0$ where

$$(3)\quad \begin{cases} E_0 = \sum_{i=0}^{k} \alpha_i \\ E_1 = \sum_{i=0}^{k} (i\alpha_i - \beta_i) \\ E_2 = \frac{1}{2!} \sum_{i=1}^{k} i^2\alpha_i - \sum_{i=1}^{k} i\beta_i - \sum_{i=0}^{k} \gamma_i \\ \vdots \\ E_q = \frac{1}{q!} \sum_{i=1}^{k} i^q\alpha_i - \frac{1}{(q-1)!} \sum_{i=1}^{k} i^{q-1}\beta_i - \frac{1}{(q-2)!} \sum_{i=1}^{k} i^{q-2}\gamma_i \\ \qquad q = 3,4,\ldots \end{cases}$$

We shall call E_{p+1} the principal truncation error coefficient for a method of order p. As noted in [7], order is a first crude measure of accuracy of formulas given by (1) and the principal truncation error coefficient should not be used as a finer measure of accuracy within the class of all formulas of a given order p, but we can use the error constant of the method which is defined by

$$(4)\qquad E = E_{p+1} / \sum_{i=0}^{k} \beta_i .$$

However, for all the formulas we shall consider the denominator of (4) is one, hence the error constant is the principal truncation error coefficient of the method.

In succeeding sections we shall have need of the following notation in connection with Vandermonde and generalized Vandermonde matrices;

$$(5)\qquad v_n(\lambda) = [1,\lambda,\lambda^2,\ldots,\lambda^{n-1}]^T$$

$$(6)\qquad v_n'(\lambda) = \frac{dv_n(\lambda)}{d\lambda} = [0,1,2\lambda,\ldots,(n-1)\lambda^{n-2}]^T .$$

Then matrix

$$(7)\qquad V_n(\lambda_1,\ldots,\lambda_t) \equiv [v_n(\lambda_1),v_n(\lambda_2),\ldots,v_n(\lambda_t)]$$

is a (n x t) Vandermonde matrix which is nonsingular when t = n and $\lambda_i, i = 1,\ldots,n$ are distinct. Similarly, matrix

$$(8)\qquad V_n'(\lambda_1,\ldots,\lambda_t) \equiv [v_n(\lambda_1),\ldots,v_n(\lambda_t),v_n'(\lambda_t)]$$

is a n x (t+1) generalized Vandermonde matrix which is nonsingular when n = t+1 and $\lambda_i, i = 1,2,\ldots,t$ are distinct.

All computations were done on a CDC 6400 in double precision and then rounded to the number of digits displayed. We use the notation 1.234 (-04) to denote 1.234×10^{-4}.

3. CONSTRUCTION OF FORMULAS

For the stability reasons cited in [4; p. 17] we shall concentrate on a subclass of the formulas given by (1). In particular, we require $\alpha_j = 0$, $j = 0,1,\ldots,k-2$, $\alpha_{k-1} = -1$, $\alpha_k = 1$, and $\gamma_j = 0$, $j = 0,1,\ldots,k-1$; thus we shall consider only second derivative formulas of the form

$$(9) \qquad y_{n+k} = y_{n+k-1} + h\,\{\sum_{i=0}^{k} \beta_i f_{n+i}\} + h^2 \gamma_k f'_{n+k}.$$

We see that (9) contains k+2 parameters, which can be determined by further restrictions put on (9). Enright in [4] and [5] chooses to require the order of the method to be as high as possible, that is, of order k+2. In effect he requires that $\beta_i i = 0,\ldots,k$ and γ_k be chosen to satisfy the system

$$(10) \qquad E_q = 0, \quad q = 1,2,\ldots,k+2.$$

This can be done for $k \geq 1$ because, a system equivalent to (10) is

$$(11) \qquad (q-1)!\, E_q = 0, \quad q = 1,2,\ldots,k+2$$

which has nonsingular coefficient matrix

$$(12) \qquad V'_{k+1}(0,1,\ldots,k).$$

Coefficients for formulas of form (9) with requirement (10) were given in [4] for k = 1,2,...,7 and are repeated here in Table 1. These formulas are best in the sense of Newton-Cotes [14; p. 218].

Table 1. Coefficients of k-step Second Derivative Formulas of Order k+2.

k	order	γ_k	β_0	β_1	β_2	β_3	β_4	β_5	β_6	β_7
1	3	$\frac{-1}{6}$	$\frac{1}{3}$	$\frac{2}{3}$						
2	4	$\frac{-1}{8}$	$\frac{-1}{48}$	$\frac{5}{12}$	$\frac{29}{48}$					
3	5	$\frac{-19}{180}$	$\frac{7}{1080}$	$\frac{-1}{20}$	$\frac{19}{40}$	$\frac{307}{540}$				
4	6	$\frac{-3}{32}$	$\frac{-17}{5760}$	$\frac{1}{45}$	$\frac{-41}{480}$	$\frac{47}{90}$	$\frac{3133}{5760}$			
5	7	$\frac{-863}{10080}$	$\frac{41}{25200}$	$\frac{-529}{40320}$	$\frac{373}{7560}$	$\frac{-1271}{10080}$	$\frac{2837}{5040}$	$\frac{317731}{604800}$		
6	8	$\frac{-275}{3456}$	$\frac{-731}{725760}$	$\frac{179}{20160}$	$\frac{-5771}{161280}$	$\frac{8131}{90720}$	$\frac{-13823}{80640}$	$\frac{12079}{20160}$	$\frac{247021}{483840}$	
7	9	$\frac{-33953}{453600}$	$\frac{8563}{12700800}$	$\frac{-35453}{5443200}$	$\frac{86791}{3024000}$	$\frac{-2797}{36288}$	$\frac{157513}{1088640}$	$\frac{-133643}{604800}$	$\frac{1147051}{1814400}$	$\frac{1758023}{3528000}$

If instead of requiring (9) to have order k+2, we require β_i i = 0,...,k and γ_k to satisfy

(13) $$(q-1)!\ E_q = 0,\ q = 1,2,\ldots,r,$$

that is, have order r, we obtain a large family of new methods. The coefficient matrix in (13) is of size r x (k+2) and is

(14) $$V_r'(0,1,2,\ldots,k) = [v_r(0),v_r(1),\ldots,v_r(k),v_r'(k)].$$

We see that (14) is of full rank r since $V_r(0,1,\ldots,r-1)$ is a submatrix of rank r.

The solution of system (13) is not unique, but since its coefficient matrix has full rank we can conveniently use the Moore-Penrose generalized inverse to characterize its solutions; see [13] and [2]. Let B denote an r x t matrix. If $r \leq t$ and B has full rank r, the Moore-Penrose generalized inverse of B, denoted B^+, is given by

(15) $$B^+ = B^*(BB^*)^{-1}$$

where B^* is the conjugate transpose of B. The solutions of (13) can be obtained from the following theorem due to Penrose [13].

THEOREM 3.1. A necessary and sufficient condition for equation $BX = K$, $B \in R_{rxt}$, $X \in R_{tx1}$ and $K \in R_{rx1}$ to have a solution is

(16) $$BB^+K = K$$

in which case the general solution is given by

(17) $$X = B^+K + (I-B^+B)\hat{Y}$$

where $\hat{Y}$ is an arbitrary t x 1 matrix.

We observe that (14) satisfies (16) since (15) holds for V_r'. Hence the general solution of (13) is given by (17).

This approach allows us to construct a phlethora of k-step formulas of order r of form (9). However, the numerous solutions available from (17), make it imperative that an alternate procedure be developed which ensures a choice for $\hat{Y}$ that leads to good methods for the solution of (2). The case in which r = k+2 degenerates to Enright's formulas of course, but of particular interest later will be the case in which r = k+1.

Before developing alternate procedures for the construction of k-step formulas of the form (9) of order k+1 we shall consider an alternate development of Enright's formulas of order k+2. This will motivate further development.

Replace (2) by an equivalent integral equation. For our purposes we shall consider the form

(18) $$y(x+h) - y(x) = \int_x^{x+h} f(s,y(s))ds.$$

In (18) we replace the integrand by an interpolant. Since we desire formulas of the form (9), choose the interpolant so that it interpolates the data given in

(19) $\{(x_{n+k}, f'(x_{n+k}, y(x_{n+k}))), (x_{n+i}, f(x_{n+i}, y(x_{n+i})))\ i=0,1,\ldots,k\}$

It will be convenient to express the interpolant in terms of the fundamental interpolants denoted by $L_{i,j}(s)$ where

$$L_{i,j}^{(r)}(x_t) = \delta_{it}\,\delta_{jr}.$$

Hence our interpolant takes the form

$$(20)\quad \sum_{i=0}^{k} f(x_{n+i}, y(x_{n+i}))L_{i,0}(s) + f'(x_{n+k}, y(x_{n+k}))L_{k,1}(s)$$

and from (18) we obtain the approximate equation

$$(21)\quad y(x_{n+k}) - y(x_{n+k-1}) \doteq \sum_{i=0}^{k} f(x_{n+i}, y(x_{n+i}))\int_{x_{n+k-1}}^{x_{n+k}} L_{i,0}(s)\,ds$$
$$+ f'(x_{n+k}, y(x_{n+k}))\int_{x_{n+k-1}}^{x_{n+k}} L_{k,1}(s)\,ds.$$

Assuming $y(x)$ is t-times differentiable in an appropriate neighborhood of x_n we can expand terms on both sides of (21) in Taylor series and equate coefficients of the terms $y^{(j)}(x_n)$. This leads to a system

$$(22)\quad \begin{cases} \displaystyle\sum_{i=0}^{k} \frac{1}{h}\int_{x_{n+k-1}}^{x_{n+k}} L_{i,0}(s)ds = k-(k-1) \\[2ex] \displaystyle\sum_{i=0}^{k} i\int_{x_{n+k-1}}^{x_{n+k}} \frac{1}{h} L_{i,0}(s)ds \\ \displaystyle\quad + \int_{x_{n+k-1}}^{x_{n+k}} \frac{1}{h^2} L_{k,1}(s)ds = \frac{k^2-(k-1)^2}{2!} \\ \quad\vdots \\ \displaystyle\sum_{i=0}^{k} \frac{i^{p-1}}{(p-1)!}\int_{x_{n+k-1}}^{x_{n+k}} \frac{1}{h} L_{i,0}(s)ds \\ \displaystyle\quad + \frac{k^{p-2}}{(p-2)!}\int_{x_{n+k-1}}^{x_{n+k}} \frac{1}{h^2} L_{k,1}(s)ds = \frac{k^p-(k-1)^p}{p!} \\ 3 \le p \le t. \end{cases}$$

In the cases t = k+2 and t = k+1 system (22) is (10) and (13) respectively, with $h\,\beta_i = \int_{x_{n+k-1}}^{x_{n+k}} L_{i,0}(s)\,ds$ i = 0,1,...,k and $h^2\gamma_k = \int_{x_{n+k-1}}^{x_{n+k}} L_{1,k}(s)ds$.

Thus any choice of an interpolant for the data in (19) leads to a formula of the form (9). We observe that the interpolation problem represented by (19) can be formulated as a Hermite-Birkhoff (H-B) problem, which has numerous solutions in term of g-splines [14] depending upon the type of restrictions we place on the interpolant.

If we require the interpolant to be a polynomial, that is, in the terminology of [14], we require the H-B problem to be k+2 poised or equivalently, require the linear functional defined by (18) to be exact for polynomials of degree k+1 or less, then we obtain a formula which is best in the sense of Newton-Cotes. Hence, it must follow that the interpolant used to generate Enright's formulas is the Hermite or osculatory polynomial [11] through the data in (19).

If, however, we allow the H-B problem represented by (19) to only be k+1 poised we obtain a class of k-step formulas of order (k+1) generated by a natural g-spline interpolant of order k+1 [14; p. 213] to the data of (19). These formulas are displayed in Table 2. We remark that these formulas represent but one solution of (13), and hence a particular choice of $\hat{Y}$ in (17). These formulas represent the best approximation to the linear functional defined by (18), of order k+1, in the sense of Sard.

Enright in [5] develops another class of k-step second derivative formulas of order k+1 which he calls modified formulas. These, of course, represent another solution of (13). He requires $\gamma_k = -(\beta_k/2)^2$ in (13), so that an iteration scheme necessary at each stage of a method based on formulas of type (9) is easier to use. These formulas are displayed in Table 3.

4. COMPARISONS OF FORMULAS

In this section we shall compare the three classes of formulas developed previously by means of principal truncation error coefficients, L-2 norms of Peano kernels [1], and stability regions. We shall use the generic names Enright's formulas, g-spline formulas and Enright's modified formulas for the formulas displayed in Tables 1,2, and 3, respectively.

For the sake of comparison by way of principal truncation error coefficients, in addition to the three classes of formulas given previously, we include the stiffly stable methods of

Table 2. Coefficients of k-step Second Derivative Formulas of Order k+1 Based on g-splines of Order k+1

k=1, order = 2

γ_1 =- 1.25(-01)
β_0 = 3.75(-01)
β_1 = 6.25(-01)

k = 2, order = 3*

γ_2 = - 1.25(-01)
β_0 = - 2.0833333333333(-02)
β_1 = 4.1666666666667(-01)
β_2 = 6.0416666666667(-01)

k = 3, order = 4

γ_3 = - 1.0792349726776(-01)
β_0 = 5.6921675774135(-03)
β_1 = - 4.6448087431693(-02)
β_2 = 4.6789617486339(-01)
β_3 = 5.7285974499089(-01)

k = 4, order = 5

γ_4 = - 9.5656639984219(-02)
β_0 = - 2.4747288928341(-03)
β_1 = 1.9680035576597(-02)
β_2 = - 7.9696746714010(-02)
β_3 = 5.1459566228535(-01)
β_4 = 5.4789577774490(-01)

k = 5, order = 6

γ_5 = - 8.7009180795033(-02)
β_0 = 1.3481638409934(-03)
β_1 = - 1.1377412895098(-02)
β_2 = 4.4691619572113(-02)
β_3 = - 1.1912076269150(-01)
β_4 = 5.5592631824706(-01)
β_5 = 5.2853207392644(-01)

k = 6, order = 7

γ_6 = - 8.0603042408652(-02)
β_0 = - 8.3533949273781(-04)
β_1 = 7.6414284746964(-03)
β_2 = - 3.1915178269140(-02)
β_3 = 8.2752204048137(-02)
β_4 = - 1.6368154701447(-01)
β_5 = 5.9296904713539(-01)
β_6 = 5.1306938511812(-01)

k = 7, order = 8

γ_7 = - 7.5635595222799(-02)
β_0 = 5.6230914806574(-04)
β_1 = - 5.5994113935069(-03)
β_2 = 2.5410857207101(-02)
β_3 = - 7.0223925398391(-02)
β_4 = 1.3554933880279(-01)
β_5 = - 2.1274589301775(-01)
β_6 = 6.2670989428697(-01)
β_7 = 5.0033683036472(-01)

* this formula is really of order 4

Table 3. Coefficients of k-step Second Derivative Formulas of Order k+1 where $\gamma_k = -(\frac{\beta_k}{2})^2$.

k	order	Z	β_0	β_1	β_2	β_3	β_4	β_5	β_6
1	2	$\sqrt{2}$	$-1 + Z$	$2 - Z$					
2	3	$\sqrt{6}$	$\frac{2}{9} - \frac{1}{9} Z$	$\frac{-5}{9} + \frac{4}{9} Z$	$\frac{4}{3} - \frac{1}{3} Z$				
3	4	$\sqrt{5}$	$\frac{-257}{2904} + \frac{6}{121} Z$	$\frac{137}{363} - \frac{27}{121} Z$	$\frac{-1103}{2904} + \frac{54}{121} Z$	$\frac{12}{11} - \frac{3}{11} Z$			
4	5	$\sqrt{1419}$	$\frac{1057}{22500} + \frac{1}{625} Z$	$\frac{-3661}{15000} - \frac{16}{1875} Z$	$\frac{3853}{7500} + \frac{12}{625} Z$	$\frac{-12449}{45000} - \frac{16}{625} Z$	$\frac{24}{25} + \frac{1}{75} Z$		
5	6	$\sqrt{5118}$	$\frac{-261979}{9009120} - \frac{10}{18769} Z$	$\frac{2416169}{13513680} + \frac{125}{37538} Z$	$\frac{-2083057}{4504560} - \frac{500}{56307} Z$	$\frac{2889973}{4504560} + \frac{250}{18769} Z$	$\frac{-5534137}{27027360} - \frac{250}{18769} Z$	$\frac{120}{137} + \frac{5}{822} Z$	
6	7	$\sqrt{117573}$	$\frac{1231883}{62233920} + \frac{5}{64827} Z$	$\frac{-20297}{144060} - \frac{4}{7230} Z$	$\frac{124541}{288120} + \frac{25}{14406} Z$	$\frac{-2887799}{3889620} - \frac{200}{64827} Z$	$\frac{587501}{768320} + \frac{25}{7203} Z$	$\frac{-10783}{72030} - \frac{20}{7203} Z$	$\frac{40}{49} + \frac{1}{882} Z$

Gear [6]. Table 4 displays these error coefficients for orders 2 through 8. In this mode of comparison, the g-spline formulas are superior.

Table 4. Principal Truncation Error Coefficients

Order	Enright's Formulas	g-spline Formulas	Enright's Modified Formulas	Gear's Stiffly Stable Formula
2	+	-.208333(-01)	-.404401(-01)	-.333333(+00)
3	.138889(-01)	.486111(-02)*	-.194066(-01)	-.25 (+00)
4	.486111(-02)	.591985(-03)	-.119245(-01)	-.2 (+00)
5	.236111(-02)	.381328(-03)	.881604(-01)	-.166667(+00)
6	.135582(-02)	.232350(-03)	.573520(-01)	-.142857(+00)
7	.863331(-03)	.147326(-03)	.404984(-01)	+
8	.589933(-03)	.979128(-04)	+	+

\+ No formula available.

* The formula for k = 2 in Table 2 is the same as that given for k = 2 in Table 1. It follows that this formula is really of order 4. In terms of the development employing the integration of an interpolant, Enright's formula uses an osculatory polynomial while this formula uses a g-spline, which is not a polynomial. It is the approximation of the linear functional (18), which yields the same formula.

A second comparison of these formulas can be made using a theorem due to Peano [3; p. 70]. If we recall that the interpolant in (20) can be viewed as a solution of a particular H-B problem, then by making the same assumptions on f we can use the development given in [1; p. 832-833] for bounding the absolute value of the functional

$$(23)\quad Rf \equiv \int_{x_{n+k-1}}^{x_{n+k}} f(s,y(s))ds - \sum_{i=0}^{k} f(x_{n+i},y(x_{n+i}))\int_{x_{n+k-1}}^{x_{n+k}} L_{i,0}(s)ds$$
$$-f'(x_{n+k},y(x_{n+k}))\int_{x_{n+k-1}}^{x_{n+k}} L_{k,1}(s)ds$$

in terms of the L^2 norm of the Peano kernel of Rf; see [1], [3]. Thus we establish that the formulas of Tables 1,2 and 3 can be compared by computing the L^2 norms of their Peano kernels and all integrations can be carried out exactly. Schoenberg [14] has shown that the L^2 norm of the Peano kernel of Rf, (23), is minimized when the interpolant used is a natural g-spline of order k+1. Thus a priori to any calculations, we know that the L^2 norms of the Peano kernels of our g-spline formulas are guaranteed to be smaller than those of the corresponding modified formula of Enright. However, the theory does not extend to enable us, a priori, to compare Enright's formulas and the g-spline formulas; see [14; eqn (8.1), p. 219]. But comparisons made in [1] of an analogous nature for natural splines and Lagrange polynomials lead us to suspect there is hope in this case. The data given in Table 5 show us that the L^2 norms of the Peano kernels of the g-spline formulas are smaller than those of the corresponding formulas of Enright.

Table 5. L^2 Norms of Peano Kernels

Order	Enright's Formulas	g-spline Formulas	Enright's Modified Formulas
2	+	.559017(-01)	.603123(-01)
3	.162650(-01)	.109487(-01)*	.200584(-01)
4	.465847(-02)	.333366(-02)	.101058(-01)
5	.194767(-02)	.136633(-02)	.615564(-01)
6	.100671(-02)	.681997(-03)	.376551(-01)
7	.593517(-03)	.387634(-03)	.253319(-01)
8	.382158(-03)	.240948(-03)	+

\+ No formula available.
* Really a fourth order Formula.

Our final comparison involves the graphs of the boundaries of the stability regions of Enright's formulas and the g-spline formulas; Fig. 1. Enright in [4] gave a sketch of stability boundaries for his formulas, together with a chart for best values of D and θ required by the definition of stiff stability. As figure 1 shows, the stability regions for both classes of formulas are similar in shape with the g-spline formulas having a slightly larger region of stability for the same value of k. Recall, however, that Enright's k-step formula is of order k+2 while a k-step g-spline formula is of order k+1.

5. SUMMARY AND CONCLUSIONS

It is possible to generate g-spline formulas of orders 2, 4,5,6,7 and 8. These formulas are stiffly stable and use y''. In the process of generating them, a unified strategy for doing so was developed.

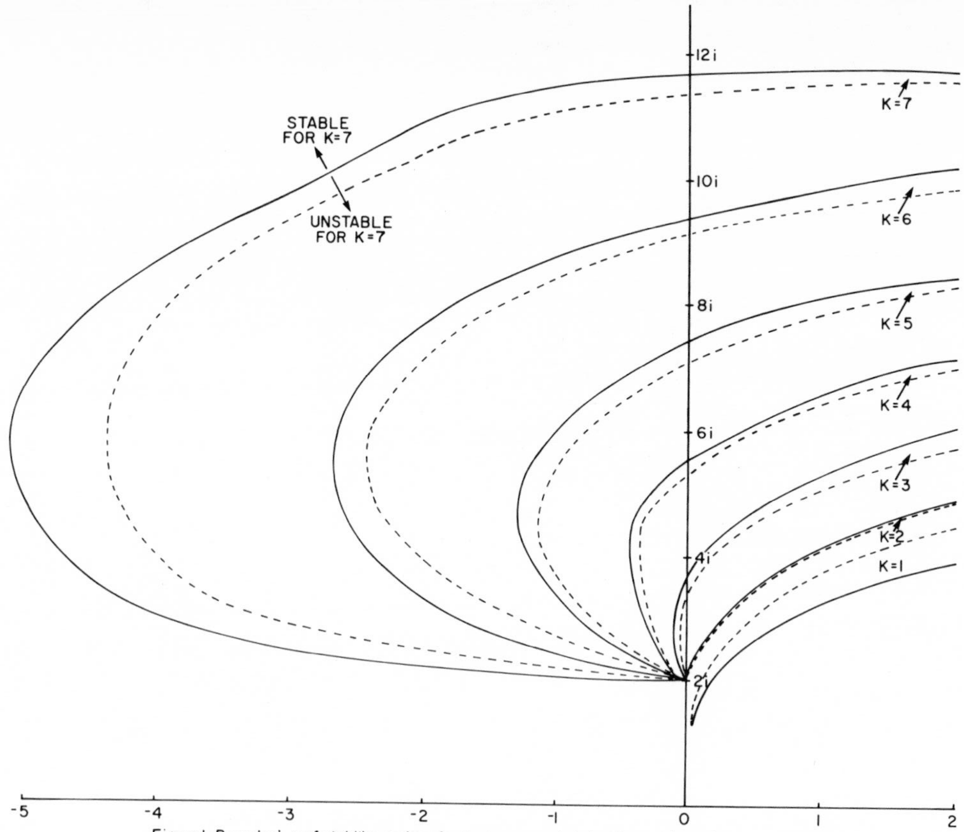

Figure 1. Boundaries of stability regions for k-step second derivative multistep formulas. Regions are symmetric about the real axis. Solid line, Enright's formulas, broken line g-spline formulas.

The local truncation error constants for a given order are smaller for a g-spline formula than for the other stiff methods considered. A similar comparison was done using the L-2 norms of the Peano kernel. In general, the comparison of stability regions is favorable to the g-spline formulas if a g-spline formula of order q is compared with another formula of order q+1.

A g-spline formula of order q uses as much of the history of the solution as an Enright formula of order q+1. Further, a g-spline formula of order q requires more work (or a special type of problem) than a comparable backward differentiation method of Gear. An underlying question that remains unsettled is "Is it feasible to use second derivative g-spline methods in a solver for stiff ordinary differential equations?"

REFERENCES

[1] Andria, G.D., Byrne, G.D., and Hill, D.R., "Integration Formulas and Schemes Based on g-splines", Math. Comp., 27 (1973), pp. 831-838 with microfiche supplement in Addendum Section.

[2] Boullion, T. and Odell, P. Generalized Inverse Matrices, John Wiley and Sons, Inc., New York, 1971.

[3] Davis, P.J., Interpolation and Approximation, Blaisdell Publishing Co., Waltham, Mass., 1963.

[4] Enright, W.H., Studies in the Numerical Solution of Stiff Ordinary Differential Equations, Ph.D. Thesis, University of Toronto, 1972.

[5] Enright, W.H., "Optimal Second Derivative Methods for Stiff Systems," Stiff Differential Systems edited by R.A. Willoughby, Plenum Press, 1974.

[6] Gear, C.W., Numerical Initial Value Problems in Ordinary Differential Equations, Prentice-Hall, Inc., Englewood Cliffs, N.J., 1971.

[7] Henrici, P., Discrete Variable Methods in Ordinary Differential Equations, John Wiley and Sons, Inc., New York, 1962.

[8] Hill, D.R., An Approach to the Numerical Solution of Delay Differential Equations, Ph.D. Thesis, University of Pittsburgh, 1973.

[9] Hill, D.R., "A New Class of One-Step Methods for the Solution of Volterra Functional Differential Equations", BIT 14 (1974), pp. 298-305.

[10] Hill, D.R., "On Comparing Adams and Natural Spline Multistep Formulas", Math. Comp., 29 (1975), pp. 741-745.

[11] Isaacson, E. and Keller, H.B., Analysis of Numerical Methods, John Wiley and Sons, Inc., New York, 1966.

[12] Lapidus, L. and Seinfeld, J.H., Numerical Solution of Ordinary Differential Equations, Academic Press, New York, 1971.

[13] Penrose, R., "A Generalized Inverse for Matrices", Proceedings of the Cambridge Philosophical Society, Vol. 51, Part 3 (1955), pp. 406-413.

[14] Schonenberg, I.J., "On the Ahlberg-Nilson Extension of Spline Interpolation: the g-spline and their Optimal Properties", J. Math. Anal. Appl. 21 (1968), pp. 207-231.

Numerical Integration of Linearized Stiff Ordinary Differential Equations

R. Leonard Brown

1. INTRODUCTION

Although numerical methods exist for the solution of stiff ordinary differential equations, most have some restriction placed on them by the fact that no numerical method requiring only first derivatives exists that is both A-stable, i.e. suitable for any stable ordinary differential equation, and of order greater than two. Thus, a variable stepsize variable order method must be restricted in stepsize or order for certain equations if it is a first derivative method, or must be restricted by requiring that higher derivatives be calculated, or both.

This paper will present a variable stepsize variable order method that is A-stable for orders up to 7 provided the second and third analytic derivatives can be computed. Methods for computing these higher derivatives for linear inhomogeneous systems of ordinary differential equations are presented; a similar treatment for equations arising in chemical kinetics is proposed. Numerical tests run on sample problems are described, showing the applicability of the treatment.

2. THE NUMERICAL METHOD

The problem to be solved involves finding numerical approximations y_n to $y(t_n)$ where $t_n = t_0 + nh$, and $y(t_n)$ is the solution to the initial value problem

$$y' = f(y,t) \tag{1a}$$

$$y(t_0) = y_0. \tag{1b}$$

It has been shown [1] that formulas of the form

$$0 = \sum_{i=0}^{k} \alpha_i^{(0)} y_{n-i} + \sum_{j=1}^{s} \alpha_0^{(j)} h^j y_n^{(j)} \tag{2}$$

are apparently A-stable for orders $N = k+s-1$, where $s = 1$ for $N \le 2$, $s = 2$ for N = 3,4, or 5, and $s = 3$ for N = 6 or 7. The formulas are called apparently A-stable since a numerical investigation of the boundary of the stability region with a very small meshsize reveals no portion of the negative complex half plane outside of the stability region.

These formulas (2) were programmed as a variable stepsize variable order method in a FORTRAN program ASTIFF, requiring a subroutine DIFFUN (T,H,Y,DYS,NORD) which places the first through NORD-th derivative elements of the Taylor expansion of Y into DYS. The dependent variables y_i are stored in Y(1,I); $h^j y_i^{(j)}/j!$ is stored in DYS (I,J), where $y_i^{(j)}$ is the J-th time derivative of Y(1,I). Rather than storing the previous computed points y_{n-i}, $i = 0, \ldots, k$, the elements of the Taylor expansion of Y are stored, allowing easy change of stepsize and order, as described in [2].

A feature found in many stiff integrators is the use of a Newton type iteration for the corrector step, requiring a matrix $W = I - \sum_{j=1}^{s} h^j \alpha_0^{(j)} \dfrac{\partial f^{(j-1)}}{\partial y}$. ASTIFF requires that a matrix PW computed with approximate or exact Jacobian terms $\dfrac{\partial f^{(j-1)}}{\partial y}$ be computed whenever a subroutine MATRIX is called. Further use can be made of the required computation of $\dfrac{\partial f}{\partial y}$.

3. APPLICATIONS

The numerical method in ASTIFF is well suited to solving the system of equations

$$y' = Ay + g(t) \tag{3}$$

where A is a matrix, g is a vector valued function of the independent variable, and y is the solution vector. The second and third derivatives can be computed using

$$y'' = Ay' + g'(t) \tag{4a}$$

$$y''' = Ay'' + g''(t) \tag{4b}$$

$$\frac{\partial f^{(j-1)}}{\partial y} = A^j \tag{4c}$$

If a numerical approximation to $g'(t)$ of order at least N, and to $g''(t)$ of order at least N-1 is used, then the error due to the numerical approximation will be of the same order as the ignored high order terms of the truncation error, since the terms actually computed are $\frac{h^2}{2}\ g'(t)$ and $\frac{h^3}{6}\ g''(t)$. See [1].

For the more general problem

$$y' = f(y) + g(t) \tag{5}$$

a similar treatment is

$$y' = J\ y + (f(y) - Jy) + g(t) \tag{6}$$

where J is an approximation to $\frac{\partial f}{\partial y}$. Higher order terms can be computed from

$$y'' = Jy' + g'(t), \tag{7a}$$

$$y''' = Jy'' + g''(t) \tag{7b}$$

where $g'(t)$ and $g''(t)$ are approximated as before, and it is assumed that

$$\frac{\partial f}{\partial y}\ y' - Jy' = O(h^N), \tag{8a}$$

$$\frac{\partial f}{\partial y}\ y'' - Jy'' = O(h^{N-1}). \tag{8b}$$

This leads to the following result:

THEOREM Given a N-th order formula (2) for which $\rho(\xi) = \sum_{i=0}^{k} \alpha_i^{(0)} \xi^{k-i}$ has all but one root inside the unit circle; and an initial value problem (5) with $f^{(j)}(y)$, $j=0,1,2$, and $g(t)$ continuous for bounded y and for $t_0 \leq t \leq t_f$; and $\frac{\partial f^{(j)}}{\partial y}$, $j=0,1,2$, exists and is bounded; and an exact initial value, then an iterated corrector method in which (2) is solved exactly with approximations (7) and J is re-evaluated frequently enough that (8) holds, is convergent, and there exist constants C_0, D, C_2, h_0 dependent only on (5) and (2) such that for $t_n \le t_f$,

$$|y_n - y(t_n)| \le C_0 D h^N \exp(C_2(t_n - t_0))$$

for any $h < h_0$.

The theorem follows as a corollary of theorem 2.1 of [1] by allowing D to bound the sum of the truncation error and the error propogated from using approximate values for y'', y''', i.e.

$$|\tilde{y}_n - y(t_n)| \le D h^{N+1},$$

where $\tilde{y}_n$ is the computed solution starting from the exact solution $y(t_n - h)$.

If $\frac{\partial f}{\partial y}$ is calculated exactly by MATRIX whenever the stepsize changes or the corrector fails to converge, this will result in (8) being satisfied. These methods were tested on several sample systems of equations. The results are in the next section.

4. TEST RESULTS

An example of equation (3) is

$$y' = \begin{pmatrix} \alpha_1 & \beta_1 & 0 & 0 \\ -\beta_1 & \alpha_1 & 0 & 0 \\ 0 & 0 & \alpha_2 & \beta_2 \\ 0 & 0 & -\beta_2 & \alpha_2 \end{pmatrix} y + \begin{pmatrix} (\delta_1-\alpha_1-\beta_1)e^{\delta_1 t}+\alpha_1+\beta_1 \\ (\delta_1-\alpha_1+\beta_1)e^{\delta_1 t}+\alpha_1-\beta_1 \\ (\delta_2-\alpha_2-\beta_2)e^{\delta_2 t}+\alpha_2+\beta_2 \\ (\delta_2-\alpha_2+\beta_2)e^{\delta_2 t}+\alpha_2-\beta_2 \end{pmatrix}$$

The correct solution is

$$y = e^{\alpha_i t}(\cos \beta_i t \pm \sin \beta_i t) + e^{\alpha_i t}, \; i = 1,2.$$

Table I lists the total number of steps, number of calls to DIFFUN, and number of calls to MATRIX for solving the above problem with

$\alpha_1 = -1,\ \beta_1 = 20,\ \delta_1 = -.5,\ \alpha_2 = -10,\ \beta_2 = 3,\ \delta_2 = -50.$

A subroutine DIFF was called to approximate $g'(t)$ and $g''(t)$ when needed. Three different error per step criteria were used to control estimated truncation error.

The test problem was integrated up to 5000 calls to DIFFUN or until T = 100., and the global error computed. A computation with exact derivatives of $g(t)$ gave the same results, with the global error staying within 10 times the truncation error per step criterion for both solutions.

TABLE I - *Test results for linear inhomogeneous equation*

ERROR	T	STEPS	DIFFUN	MATRIX
1.e-4	1.e 2	1345	2164	476
1.e-6	1.e 2	1853	3048	475
1.e-8	1.8 e 1	2468	5003	61

A simple form of (5) is attributed to Robertson and described in [3] as

$$y_1 = -k_1\, y_1 + k_2\, y_2\, y_3 \qquad y_1(0) = 1$$

$$y_2 = k_1\, y_1 - k_2\, y_2\, y_3 - k_3\, y_2^2 \qquad y_2(0) = 0$$

$$y_3 = k_3\, y_2^2 \qquad y_3(0) = 0$$

where k_i are rate constants and y_i are reaction component concentrations. The lack of dependence on t implies $g(t) = 0$. The problem was run twice at two error per step criteria, once using exact second derivatives and restricting the calculation to order 5 (note that $y'' = \frac{\partial f}{\partial y}\, y'$ for this autonomous system and $\frac{\partial f}{\partial y}$ is computed exactly at every call to DIFFUN) and again using the approximation (7). The numerical solutions were the same to four decimal places. Table II compares the runs using the same output format as above, with $k_1 = 0.04$, $k_2 = 1.e\ 4$, $k_3 = 3.e\ 7$.

TABLE IIA - *Test results for non-linear equation; exact* y''

ERROR	T	STEPS	DIFFUN	MATRIX
1.e-6	1.e 3	628	2229	543
1.e-7	1.e 3	1365	4261	1059

TABLE IIB - *Test results for non-linear equation; Approximate* y'', y'''

ERROR	T	STEPS	DIFFUN	MATRIX
1.e-6	1.e 3	504	2047	456
1.3-7	1.e 3	1164	3945	874

5. CONCLUSIONS

By proper linearization of certain systems of ordinary differential equations, very good approximations to y'' and y''' can be made and thus ASTIFF can be used to integrate any such system if it is analytically stable, with no restraint on the stepsize other than controlling the truncation error. Time dependent forcing terms can be easily accounted for by using numerical differentiation of the non-autonomous terms. ASTIFF can be a useful tool in integrating equations of the forms (3) and (5).

REFERENCES

1. Brown, R.L. (1974) "Multi-derivative numerical methods for the solution of stiff ordinary differential equations," Report UIUCDCS-R-74-672, Univ. of Illinois Dept. of Computer Science.
2. Gear, C.W. (1971) "DIFSUB for solution to ordinary differential equations," *Comm.ACM*, *14*, pp. 185-190.
3. Edsberg, L. (1974) "Integration package for chemical kinetics," *Stiff Differential Equations*, ed. R. Willoughby, pp. 81-94, Plenum Press.

Comparing Numerical Methods for the Solution of Stiff Systems of ODEs Arising in Chemistry

W. H. Enright and T. E. Hull

1. INTRODUCTION

In chemical applications one often encounters systems of ordinary differential equations which, although mathematically well conditioned, are virtually impossible to solve with traditional numerical methods because of the severe stepsize constraint imposed by numerical stability. These stiff equations can be characterized by the presence of transient components which, although negligible relative to the other components of the numerical solution, constrain the stepsize of traditional numerical methods to be of the order of the smallest time constant of the problem.

One of the first attempts to cope with the difficulties of 'stiffness' was suggested by Curtiss and Hirschfelder (1952), who encountered stiff equations in their kinetics studies. They proposed special multistep formulas which were able to produce acceptable approximations without the severe stepsize constraint imposed by stability for the more traditional Adams multistep formulas and Runge-Kutta formulas.

The difficulty of stiffness was virtually ignored by numerical analysts until ten years later when Dahlquist (1963) identified numerical instability as

---

This work was supported in part by the National Research Council of Canada and by USERDA while the first author held a visiting appointment at Argonne National Laboratory.

the cause of the difficulty and provided some basic definitions and concepts that have been very helpful in subsequent work. In the same paper, Dahlquist also proposed the trapezoidal rule with extrapolation as a suitable technique for integrating stiff equations. Since Dahlquist's paper appeared, the field has been quite active and several new approaches have been proposed for the numerical solution of stiff equations. In addition, several survey papers discussing the difficulty of stiffness are available (Cooper (1969), Bjurel et al (1970), Seinfeld, Lapidus and Hwang (1970), and Gelinas (1972)).

Before discussing some of the new techniques that have been proposed for the solution of stiff equations, we would like to distinguish between a numerical method and a basic formula. Most numerical methods generate approximations to the solution y(x) of the initial value problem:

$$y' = f(x,y),\ y(x_0) = y_0 \text{ over } [x_0,x_f], \tag{1}$$

at a discrete set of points $x_0, x_1 \ldots x_N = x_f$. The basic formula is used to determine y_{i+1}, the approximation to $y(x_{i+1})$, given the previously accepted approximations $(x_0,y_0), (x_1,y_1) \ldots (x_i,y_i)$ and the stepsize $h_i = (x_{i+1} - x_i)$. On the other hand, by numerical method we will mean a complete algorithm that chooses the current stepsize, applies a basic formula and monitors the error introduced on each step in an attempt to ensure that a specified error tolerance is satisfied.

Some of the more readily available methods for stiff equations include:

a) Variable-order methods based on backward differentiation multistep formulas, originally analysed and implemented by Gear (1969, 1971a, 1971b) and later modified and studied by Hindmarsh (1974) and Byrne and Hindmarsh (1975).
b) Methods based on the trapezoidal rule, such as those proposed by Dahlquist (1963) and subsequently studied by Lindberg (1971, 1972).
c) Implicit Runge-Kutta methods suitable for stiff equations, such as those based on the formulas of Butcher (1964) and studied by Ehle (1968).
d) Methods based on the use of preliminary mathematical transformations to remove stiffness and the solution of the transformed problem by traditional techniques, such as those studied and

implemented by Lawson and Ehle (1972).

e) Methods based on second derivative multistep formulas, such as those developed by Liniger and Willoughby (1967) and Enright (1974).

Unfortunately, although a number of methods have been developed, and many more basic formulas suggested for stiff equations, until recently there has been little advice or guidance to help a practitioner choose a good method for his problem. This difficulty is complicated by the fact that papers suggesting new techniques for ODE's are very difficult to referee, and often new techniques are published which are not competitive with those already available. There is a definite need for the establishment of minimum standards that must be met before a new technique is published. Fixed-stepsize comparisons or out-performance of a standard fourth order Runge-Kutta method should no longer justify the publication of a new technique for stiff equations. It is hoped that the development of testing programs such as the one used in this paper will provide a useful tool for assessing new algorithms for stiff equations.

In the next section we describe the comparisons we have performed at Toronto and we outline a new testing program which is then used to assess the performance of numerical methods in the solution of some stiff ODE's arising in chemistry. In the last section we present the results of these tests and make some specific recommendations.

2. TESTING NUMERICAL METHODS

We have been testing and comparing methods for both stiff and non-stiff ODE's at Toronto for several years (Hull et al (1972), Enright et al (1974, 1975)). Our primary purpose in these comparisons has been the identification of good methods for implementation in a software package for ODE's. In these studies we proposed some precise, computable measures of reliability and efficiency. In addition, in order to make the comparisons meaningful, we modified the error control of most of the methods tested so that all methods were attempting to perform the same task. Although we feel that our modifications did not appreciably affect the relative performances of the methods tested, it is true that our tests were on modified programs, and therefore cannot be considered to be certifications of particular codes. Partly because of this, we have recently been interested in the development of aids for the assessment and certification of individual unmodified methods.

We have developed a new testing program intended to help in the exhaustive testing of a particular method. This exhaustive testing is necessary in the certification of a method and in the identification of the problem domain over which a method is suitable. Because of the requirement that the testing program assess unmodified programs (and the implicit assumption that the testing program not affect the method's performance), comparisons will be more difficult than in our earlier tests (where all methods were attempting to perform the same task). Nevertheless, as we shall see, our results will allow us to endorse some methods and reject others.

In this investigation we are particularly concerned with the assessment of a method's performance on stiff equations arising in chemistry. The specific test problems have been obtained from real examples in the literature and through private communications. These include standard kinetic problems, kinetic problems with oscillatory solutions and reactor problems. The test problems are specified in detail in the appendix.

The testing program monitors the performance of a numerical method in its solution of the test problems and generates measures of cost and reliability which can then be used to assess the methods. The testing program assumes that the method being tested is attempting to keep the magnitude of the global error within the specified error tolerance, TOL. It is further assumed that this control of global error is accomplished indirectly by monitoring and controlling the local error introduced on each step of the calculation.

That is, if

i) $z(x,x_i,y_i)$ is the 'local' solution of the initial value problem $z'=f(z,x)$, $z(x_i)=y_i$;

ii) ERREST is the method's estimate of the local error;

and iii) ERRLOC is the local error bound that is assumed satisfied on the ith step;

then the method has required that

ERREST $\leq$ ERRLOC,

in an attempt to ensure that the local error satisfies

$$||z(x_{i+1}, x_i, y_i) - y_{i+1}|| \leq \text{ERRLOC} .$$

The testing program measures the local and global errors on each step of the calculation and outputs the following statistics after the completion of each test problem:

- TIME - the time required to solve the problem (in seconds on an IBM 370/195).
- FCN CALLS - the number of function evaluations required to solve the problem,
- JAC CALLS - the number of Jacobian evaluations required to solve the problem,
- INV CALLS - the number of matrix inversions required to solve the problem (for methods which use LU decompositions rather than matrix inversions this is a count of the number of LU decompositions),
- NO OF STEPS - the number of steps required to solve the problem,
- MAX LOC ERROR - the maximum ratio of true local error to ERRLOC,
- FRACTION DECV - the fraction of steps where the true local error exceeded ERRLOC,
- GL ERR XEND - the magnitude of the global error at x_f, the endpoint of the integration (in units of the requested tolerance, TOL),
- GL ERR OVERALL - the maximum observed global error over all steps in the solution of the problem (in units of TOL).

Summary statistics are also output after the solution of all problems at a specified tolerance and after the completion of all problems at all tolerances. For these tests error tolerances of 10^{-2}, 10^{-4} and 10^{-6} are used.

The specific methods that we assess in this study are:

a) three methods based on backward differentiation formulas, DIFSUB (Gear (1971a, 1971b)), GEAR.REV3 (Hindmarsh (1974)) and EPISODE (Byrne and Hindmarsh (1975)),
b) our version of a method based on the trapezoidal rule with extrapolation, TRAPEX (discussed in Enright et al (1975)),
c) our version of a method based on the fourth order implicit Runge-Kutta formula of Butcher (1964), IMPRK (discussed in Enright et al (1975)),
d) a generalized Runge-Kutta method, GENRK (Lawson and Ehle (1972)),
e) a variable-order second derivative multistep method, SDBASIC (Enright (1974)).

NOTE:

a) The methods GEAR.REV3 and EPISODE use the more efficient LU decomposition followed by back substitution, rather than matrix inversions followed by matrix-vector multiplications. For these methods, the count reported as INV CALLS is actually a count of the number of LU decompositions.

b) The error control used by all of the methods reported here is based on a control of the error-per-step. (A discussion of other possible error control strategies is included in Enright et al (1975).)

3. RESULTS AND RECOMMENDATIONS

The detailed results are presented in Tables 1 through 7. In order to reduce the expense of testing some of the methods, a maximum of 15,000 function evaluations was allowed for each problem. If the method was unable to solve the problem with fewer function evaluations, the integration was aborted. (See for example the results of TRAPEX at 10^{-6} on CHM 9 and CHM 10.) On other problems the method itself has detected some difficulty and abandoned the integration. This can happen for example when solving kinetics problems at relaxed error tolerances; negative concentrations may be introduced which cause the problem to become mathematically unstable. (See the results for EPISODE at 10^{-2} on CHM 1 and CHM 3.) It can also happen at more severe error tolerances if the method cannot achieve the desired accuracy. (See the results of GENRK at 10^{-6} on CHM 6.)

In addition to these failures of the method being tested there are also several cases where the method has successfully integrated the problem, but the testing program was unable to calculate the true local solution or the true global solution throughout the range of integration. This occurs because the 'true' local and global solutions are computed numerically at an accuracy of one percent of the current tolerance and these integrations can fail, especially at the severe error tolerances. (In our testing we have used different methods for the calculation of the 'true' solutions, and only minor discrepancies in the results were observed. In the results reported here a modified version of SDBASIC was used.) When this failure of the true solution has occurred, the corresponding statistics are flagged with an asterisk (*). It should be noted that in these cases the generated statistics are still meaningful, although they are not

TABLE 1 : DIFSUB

DIFSUB		TIME	FCN CALLS	JAC CALLS	INV CALLS	NO OF STEPS	MAX LOC ERROR	FRACTION DECV	GL ERR XEND	GL ERR OVERALL
10**-2	CHM 1	.003	12	2	2	8	0.9	0.0	1.65	1.65
	CHM 2	.021	151	14	14	50	0.7	0.0	0.06	1.50
	CHM 3	.027	131	18	18	51	0.7	0.0	0.01	1.13
	CHM 4	.005	34	7	7	11	0.5	0.0	0.74	0.91
	CHM 5	.008	82	16	16	18	0.3	0.0	0.56	0.63
	CHM 6	.072	345	34	34	132	0.7	0.0	0.01	0.08
	CHM 7	.006	43	8	8	21	0.6	0.0	0.34	0.70
	CHM 8	.028	194	29	29	40	0.6	0.0	0.16	0.20
	CHM 9	.241	1743	101	101	534	2.3	0.047	0.86	35.57
	CHM 10	.014	58	13	13	22	0.0	0.0	0.01	0.03
	SUMMARY	.423	2793	242	242	887	2.3	0.028	1.65	35.57
10**-4	CHM 1	.010	50	10	10	13	0.2*	0.0*	--*	0.28*
	CHM 2	.046	270	19	19	107	2.2	0.009	0.01	3.48
	CHM 3	.066	327	16	16	137	0.7	0.0	0.02	1.36
	CHM 4	.008	45	8	8	18	0.6	0.0	0.51	1.78
	CHM 5	.015	160	20	20	39	0.6	0.0	5.68	6.16
	CHM 6	.167	679	42	42	265	1.1	0.004	0.01	0.31
	CHM 7	.014	135	11	11	56	0.7	0.0	1.06	1.45
	CHM 8	.040	322	34	34	86	0.6	0.0	0.27	0.50
	CHM 9	.338	2615	117	117	820	1.1*	0.060*	--*	7.91*
	CHM 10	.019	75	12	12	35	0.2	0.0	0.01	0.31
	SUMMARY	.724	4678	289	289	1576	2.2	0.001	5.68	7.9
10**-6	CHM 1	.009	44	7	7	15	1.3	0.067	4.86	4.86
	CHM 2	.060	440	21	21	180	1.4	0.017	0.07	4.42
	CHM 3	.104	589	28	28	246	0.7*	0.0*	--*	1.43*
	CHM 4	.011	76	8	8	31	0.6	0.0	1.67	1.67
	CHM 5	.032	334	29	29	100	0.8	0.0	14.95	15.93
	CHM 6	.200	1091	49	49	411	--*	--*	--*	--*
	CHM 7	.027	238	14	14	106	1.3	0.009	0.12	1.78
	CHM 8	.072	576	44	44	169	0.6	0.0	0.39	1.88
	CHM 9	.527	4186	118	118	1392	1.1*	0.046*	--*	8.35*
	CHM 10	.027	155	14	14	66	5.2*	1.00*	--*	5.17*
	SUMMARY	1.070	7729	332	332	2716	5.2	0.007	14.95	15.93
OVERALL	SUMMARY	2.218	15200	863	863	5179	5.2	0.009	14.95	35.57

--* statistics are missing because of failure of true solution on first step.

* statistics are a summary of all steps prior to a failure of the true solution.

TABLE 2 : GEAR.REV3

GEAR.REV3		TIME	FCN CALLS	JAC CALLS	INV CALLS	NO OF STEPS	MAX LOC ERROR	FRACTION DECV	GL ERR XEND	GL ERR OVERALL
10**-2	CHM 1	.004	15	6	6	14	0.0	0.0	0.01	0.01
	CHM 2	.013	68	13	13	48	1.2	0.042	0.07	2.10
	CHM 3	.015	77	15	15	52	1.1	0.019	0.01	2.02
	CHM 4	.003	15	5	5	14	0.1	0.0	0.01	0.07
	CHM 5	.005	33	10	10	25	0.4	0.0	0.26	0.26
	CHM 6	.048	229	25	25	129	2.3	0.047	0.01	0.21
	CHM 7	.005	34	6	6	24	0.2	0.0	0.43	0.46
	CHM 8	.014	103	12	12	40	1.0	0.0	0.38	0.38
	CHM 9	.179	928	75	75	557	10.5	0.257	1.67	127.37
	CHM 10	.013	67	20	20	46	2.9*	0.143*	--*	3.01*
	SUMMARY	.301	1569	187	187	949	10.5	0.160	1.67	127.37
10**-4	CHM 1	.004	17	6	6	16	0.0	0.0	0.01	0.01
	CHM 2	.032	177	20	20	116	1.8	0.060	0.09	7.03
	CHM 3	.036	160	21	21	116	1.5	0.060	0.09	4.35
	CHM 4	.005	22	7	7	20	0.1	0.0	0.40	0.47
	CHM 5	.014	109	17	17	51	1.0	0.0	12.65	12.86
	CHM 6	.091	416	39	39	244	2.3	0.053	0.01	0.72
	CHM 7	.013	84	9	9	54	1.3	0.019	1.02	2.76
	CHM 8	.030	200	14	14	91	1.2	0.033	0.38	2.30
	CHM 9	.299	1449	105	105	1005	2.0*	0.373*	--*	11.66*
	CHM 10	.012	52	14	14	41	0.5	0.0	0.01	0.64
	SUMMARY	.535	2686	252	252	1754	2.3	0.018	12.65	12.86
10**-6	CHM 1	.006	26	7	7	20	0.9	0.0	1.67	1.98
	CHM 2	.053	268	21	21	182	1.4	0.06	0.13	5.50
	CHM 3	.071	281	32	32	234	1.6*	0.5*	--*	4.81*
	CHM 4	.009	46	9	9	31	1.2	0.032	2.15	2.17
	CHM 5	.030	201	21	21	111	1.1	0.036	16.03	16.74
	CHM 6	.159	700	45	45	431	--*	--*	--*	--*
	CHM 7	.026	152	15	15	108	1.0	0.037	3.29	3.74
	CHM 8	.051	290	17	17	146	1.0	0.007	0.27	2.56
	CHM 9	.557	2577	162	162	1983	2.0*	0.463*	--*	19.87*
	CHM 10	.024	109	18	18	80	--*	--*	--*	--*
	SUMMARY	.985	4650	347	347	3326	2.0	0.008	16.03	19.87
OVERALL SUMMARY		1.821	8905	786	786	6029	10.5	.035	16.03	127.37

--* statistics are missing because of failure of true solution on first step.

* statistics are a summary of all steps prior to a failure of the true solution.

TABLE 3 : EPISODE

EPISODE		TIME	FCN CALLS	JAC CALLS	INV CALLS	NO OF STEPS	MAX LOC ERROR	FRACTION DECV	GL ERR XEND	GL ERR OVERALL
10**-2	CHM 1	METHOD FAILED								
	CHM 2	.016	61	20	20	41	1.4	0.098	0.03	2.07
	CHM 3	METHOD FAILED								
	CHM 4	.002	9	6	6	6	0.3	0.0	0.59	0.60
	CHM 5	.006	27	16	16	17	0.6	0.0	0.79	0.87
	CHM 6	.059	208	51	51	113	3.0	0.195	0.01	0.15
	CHM 7	.008	35	14	14	23	0.5	0.0	0.26	0.49
	CHM 8	.014	64	15	15	28	1.5	0.179	0.31	0.31
	CHM 9	.227	947	132	132	484	8.9	0.366	1.61	283.12
	CHM 10	.019	57	40	40	43	2.7	0.023	0.01	2.72
	SUMMARY	.351	1408	294	294	755	8.9	0.277	1.61	283.12
10**-4	CHM 1	.002	6	2	2	4	0.3	0.0	0.12	0.18
	CHM 2	.037	144	31	31	90	3.0	0.133	0.07	3.48
	CHM 3	.046	154	38	38	102	2.1	0.137	0.01	7.19
	CHM 4	.005	21	9	9	14	0.6	0.0	1.22	1.88
	CHM 5	.024	104	32	32	63	3.7	0.048	5.53	7.26
	CHM 6	.112	385	66	66	222	4.0	0.126	0.01	1.24
	CHM 7	.022	96	21	21	61	0.8	0.0	1.03	1.90
	CHM 8	.033	146	26	26	73	4.0	0.233	0.58	3.96
	CHM 9	.415	1643	199	199	913	7.0*	0.688*	--*	29.27*
	CHM 10	.023	65	44	44	54	0.7	0.0	0.01	0.86
	SUMMARY	.719	2764	468	468	1596	7.0	0.046	5.53	29.27
10**-6	CHM 1	.007	17	4	4	10	0.8	0.0	5.87	5.87
	CHM 2	.090	306	42	42	183	3.0	0.098	0.06	6.49
	CHM 3	.094	284	47	47	188	2.6	0.106	0.02	6.65
	CHM 4	.012	41	16	16	30	0.5	0.0	2.41	2.50
	CHM 5	.045	196	43	43	111	3.0	0.198	20.85	20.99
	CHM 6	.250	799	94	94	459	--*	--*	--*	--*
	CHM 7	.042	148	26	26	92	2.5	0.130	5.17	6.20
	CHM 8	.067	259	33	33	133	3.1	0.271	1.09	4.27
	CHM 9	.856	3003	267	267	1782	4.3*	0.750*	--*	40.76*
	CHM 10	.054	150	66	66	103	--*	--*	--*	--*
	SUMMARY	1.517	5203	638	638	3091	4.3	0.035	20.85	40.76
OVERALL SUMMARY		2.587	9375	1400	1400	5442	8.9	.071	20.85	283.1

--* statistics are missing because of failure of true solution on first step.

* statistics are a summary of all steps prior to a failure of the true solution.

TABLE 4 : TRAPEX

TRAPEX		TIME	FCN CALLS	JAC CALLS	INV CALLS	NO OF STEPS	MAX LOC ERROR	FRACTION DECV	GL ERR XEND	GL ERR OVERALL
10**-2	CHM 1	.008	4	12	12	4	0.0	0.0	0.01	0.01
	CHM 2	.030	322	30	30	15	0.0	0.0	0.01	0.01
	CHM 3	.035	234	25	25	17	0.2	0.0	0.06	0.17
	CHM 4	.143	1303	159	159	88	4.0	0.170	5.82	7.03
	CHM 5	.014	140	27	27	17	0.1	0.0	0.06	0.15
	CHM 6	.154	1017	107	107	47	0.2	0.0	0.01	0.01
	CHM 7	.006	57	11	11	6	0.0	0.0	0.07	0.07
	CHM 8	.059	653	64	64	20	0.1	0.0	0.01	0.08
	CHM 9	.482	5950	301	301	195	15.1	0.097	0.08	10.04
	CHM 10	METHOD	FAILED							
	SUMMARY	.932	9680	736	736	409	15.1	0.083	5.82	10.04
10**-4	CHM 1	.012	29	16	16	6	0.0	0.0	0.02	0.02
	CHM 2	.059	718	38	38	28	0.3	0.0	0.01	0.32
	CHM 3	.068	650	25	25	29	0.3	0.0	0.04	0.35
	CHM 4	.015	104	21	21	9	0.1	0.0	0.11	0.11
	CHM 5	.032	446	58	58	19	0.0	0.0	0.04	0.07
	CHM 6	.329	2320	175	175	91	0.6*	0.0*	--*	0.12*
	CHM 7	.011	122	16	16	11	0.1	0.0	0.01	0.14
	CHM 8	.129	1397	116	116	37	0.1	0.0	0.01	0.08
	CHM 9	1.080	13786	431	431	436	3.0	0.090	1.30	1.74
	CHM 10	METHOD	FAILED							
	SUMMARY	1.735	19572	896	896	666	3.0	0.059	1.30	1.74
10**-6	CHM 1	.021	86	22	22	9	--*	--*	--*	--*
	CHM 2	.122	1653	47	47	55	0.2	0.0	0.02	0.22
	CHM 3	.152	1669	26	26	64	--*	--*	--*	--*
	CHM 4	.018	163	22	22	9	0.0	0.0	0.01	0.02
	CHM 5	.076	1174	114	114	39	0.3	0.0	0.03	0.34
	CHM 6	.694	5554	307	307	204	--*	--*	--*	--*
	CHM 7	.023	339	15	15	23	1.0	0.0	0.01	0.98
	CHM 8	.244	3132	183	183	80	0.4	0.0	0.01	0.11
	CHM 9	METHOD	FAILED	(>15000	FCN EVALUATIONS)					
	CHM 10	METHOD	FAILED	(>15000	FCN EVALUATIONS)					
	SUMMARY	1.352	13770	736	736	483	1.0	0.0	0.03	0.98
OVERALL SUMMARY		4.019	43022	2368	2368	1558	15.1	0.047	5.82	10.04

--* statistics are missing because of failure of true solution on first step.

* statistics are a summary of all steps prior to a failure of the true solution.

TABLE 5 : IMPRK

IMPRK		TIME	FCN CALLS	JAC CALLS	INV CALLS	NO OF STEPS	MAX LOC ERROR	FRACTION DECV	GL ERR XEND	GL ERR OVERALL
10**-2	CHM 1	METHOD	FAILED							
	CHM 2	.147	341	72	72	11	0.1	0.0	0.14	0.14
	CHM 3	1.310	1257	375	375	28	0.1	0.0	0.09	0.12
	CHM 4	.592	947	299	299	17	0.2	0.0	0.21	0.23
	CHM 5	1.263	2535	790	790	38	0.1	0.0	0.02	0.12
	CHM 6	METHOD	FAILED							
	CHM 7	.010	69	9	9	6	0.0	0.0	0.01	0.03
	CHM 8	.096	367	48	48	11	4.6	0.455	0.17	0.68
	CHM 9	1.094	6819	381	381	178	2.7*	0.050*	--*	0.78*
	CHM 10	METHOD	FAILED							
	SUMMARY	4.513	12335	1974	1974	289	4.6	0.045	0.21	0.78
10**-4	CHM 1	METHOD	FAILED							
	CHM 2	.072	529	23	23	22	1.8	0.273	1.76	1.76
	CHM 3	.114	477	26	26	20	0.3	0.0	0.01	0.27
	CHM 4	.032	99	20	20	6	0.0	0.0	0.01	0.01
	CHM 5	.102	529	115	115	14	0.9	0.0	0.91	0.91
	CHM 6	.632	2141	150	150	73	0.9*	0.0*	--*	0.27*
	CHM 7	.014	133	11	11	9	0.3	0.0	0.22	0.34
	CHM 8	.141	917	54	54	33	14.7	0.667	0.01	5.98
	CHM 9	1.478	12543	305	305	433	1.3*	0.063*	--*	0.65*
	CHM 10	METHOD	FAILED	(>15000	FCN	EVALUATIONS)				
	SUMMARY	2.586	17368	704	704	610	14.7	0.035	1.76	5.98
10**-6	CHM 1	METHOD	FAILED							
	CHM 2	.201	1849	26	26	71	5.1*	0.672*	--*	5.05*
	CHM 3	.169	963	23	23	38	--*	--*	--*	--*
	CHM 4	.028	113	14	14	7	0.0	0.0	0.01	0.02
	CHM 5	.085	723	80	80	22	3.1	0.227	0.55	3.32
	CHM 6	.998	4487	161	161	188	--*	--*	--*	--*
	CHM 7	.027	311	12	12	19	0.8	0.0	0.86	1.43
	CHM 8	.369	3647	47	47	133	20.2	0.571	0.01	17.54
	CHM 9	METHOD	FAILED	(>15000	FCN	EVALUATIONS)				
	CHM 10	METHOD	FAILED	(>15000	FCN	EVALUATIONS)				
	SUMMARY	1.878	12093	363	363	478	20.2	0.267	0.86	17.54
OVERALL SUMMARY		8.977	41796	3041	3041	1377	20.2	9.118	1.76	17.54

--* statistics are missing because of failure of true solution on first step.

* statistics are a summary of all steps prior to a failure of the true solution.

TABLE 6 : GENRK

GENRK		TIME	FCN CALLS	JAC CALLS	INV CALLS	NO OF STEPS	MAX LOC ERROR	FRACTION DECV	GL ERR XEND	GL ERR OVERALL
10**-2	CHM 1	.052	142	13	39	12	0.0	0.0	0.03	0.03
	CHM 2	.025	121	11	33	11	1.1	0.364	0.07	0.13
	CHM 3	.039	154	9	27	14	7.2	0.286	0.01	7.24
	CHM 4	.012	55	5	15	5	0.0	0.0	0.02	0.02
	CHM 5	.008	66	6	18	6	1.2	0.167	0.98	0.98
	CHM 6	METHOD	FAILED							
	CHM 7	.007	66	6	18	6	0.0	0.0	0.06	0.07
	CHM 8	.029	139	13	39	9	28.7	0.667	0.58	11.83
	CHM 9	.378	2183	154	462	143	72.6	0.434	3.84	28.69
	CHM 10	METHOD	FAILED	(>15000 FCN EVALUATIONS)						
	SUMMARY	.550	2926	217	651	205	72.6	0.375	3.84	28.69
10**-4	CHM 1	.044	119	11	33	9	0.9	0.0	1.96	1.96
	CHM 2	.046	220	20	60	20	4.4	0.60	0.62	1.15
	CHM 3	.072	217	19	57	17	10.8	0.059	0.01	10.85
	CHM 4	.014	66	6	18	6	1.3	0.167	1.16	1.22
	CHM 5	.009	77	7	21	7	115.6	0.143	115.08	115.08
	CHM 6	.252	1316	43	129	116	17.5	0.241	0.01	0.94
	CHM 7	.010	88	8	24	8	0.5	0.0	0.71	0.85
	CHM 8	.082	493	33	99	33	16.2	0.909	0.42	16.21
	CHM 9	.827	5668	301	903	388	4.9*	0.379*	--*	8.62*
	CHM 10	METHOD	FAILED	(>15000 FCN EVALUATIONS)						
	SUMMARY	1.355	8264	448	1344	604	115.6	0.340	115.08	115.08
10**-6	CHM 1	.036	165	7	21	15	3.8	0.067	45.65	45.65
	CHM 2	.124	711	47	141	51	8.2	0.725	2.54	5.46
	CHM 3	.152	427	38	114	37	0.8	0.0	0.01	1.10
	CHM 4	.021	88	8	24	8	4.6	0.500	9.31	9.31
	CHM 5	.031	342	20	60	22	34.6	0.500	66.21	88.83
	CHM 6	METHOD	FAILED							
	CHM 7	.019	186	12	36	16	1.5	0.063	1.17	1.58
	CHM 8	.376	2168	138	414	138	12.8	0.971	0.35	12.90
	CHM 9	METHOD	FAILED	(>15000 FCN EVALUATIONS)						
	CHM 10	METHOD	FAILED	(>15000 FCN EVALUATIONS)						
	SUMMARY	.760	4087	270	810	287	34.6	0.655	66.21	88.83
OVERALL SUMMARY		2.665	15277	935	2805	1096	115.6	0.428	115.08	115.08

--* statistics are missing because of failure of true solution on first step.

* statistics are a summary of all steps prior to a failure of the true solution.

TABLE 7 : SDBASIC

SDBASIC		TIME	FCN CALLS	JAC CALLS	INV CALLS	NO OF STEPS	MAX LOC ERROR	FRACTION DECV	GL ERR XEND	GL ERR OVERALL
10**-2	CHM 1	.032	2	1	26	24	0.0	0.0	0.01	0.01
	CHM 2	.063	270	198	89	33	1.0	0.0	0.09	0.34
	CHM 3	.036	104	51	22	24	0.3	0.0	0.01	0.60
	CHM 4	.021	57	28	29	18	0.2	0.0	0.29	0.34
	CHM 5	.026	159	91	66	28	1.2	0.036	1.12	1.91
	CHM 6	.161	453	364	127	57	0.5	0.0	0.01	0.01
	CHM 7	.010	51	32	16	14	0.0	0.0	0.01	0.01
	CHM 8	.052	252	198	71	24	0.8	0.0	0.03	0.58
	CHM 9	.476	3055	2304	327	263	2.2	0.015	0.66	7.90
	CHM 10	.411	660	207	305	311	0.9*	0.0*	--*	1.04*
	SUMMARY	1.288	5063	3474	1078	796	2.2	0.006	1.12	7.9
10**-4	CHM 1	.033	14	1	26	24	0.0	0.0	0.16	0.16
	CHM 2	.084	402	321	100	41	0.3	0.0	0.01	0.34
	CHM 3	.070	318	206	24	46	0.7	0.0	0.03	0.70
	CHM 4	.032	133	87	36	23	0.9	0.0	3.89	3.91
	CHM 5	.042	261	218	109	27	1.7	0.037	3.40	3.79
	CHM 6	.236	762	596	142	91	0.8	0.0	0.01	0.16
	CHM 7	.018	119	85	17	25	0.2	0.0	0.06	0.22
	CHM 8	.072	400	330	82	29	1.2	0.034	0.08	0.33
	CHM 9	.763	5517	4410	448	414	1.0*	0.010*	--*	1.14*
	CHM 10	.273	364	102	173	260	1.5	0.008	0.26	2.48
	SUMMARY	1.622	8290	6356	1157	980	1.7	0.004	3.89	3.91
10**-6	CHM 1	.181	378	269	145	75	1.2	0.040	7.95	10.78
	CHM 2	.150	815	642	124	85	0.2	0.0	0.01	0.29
	CHM 3	.155	822	576	32	92	--*	--*	--*	--*
	CHM 4	.037	138	113	40	20	0.8	0.0	1.80	1.80
	CHM 5	.061	412	330	108	50	0.8	0.0	3.07	3.86
	CHM 6	.364	1306	1042	193	133	--*	--*	--*	--*
	CHM 7	.031	259	169	19	42	0.4	0.0	0.45	0.49
	CHM 8	.158	1009	832	134	68	1.2	0.015	0.23	0.75
	CHM 9	1.445	9951	8194	784	730	0.9*	0.0*	--*	1.37*
	CHM 10	.359	568	250	245	288	--*	--*	--*	--*
	SUMMARY	2.942	15658	12417	1824	1583	1.2	0.003	7.95	10.78
OVERALL SUMMARY		5.852	29011	22247	4059	3359	2.2	0.004	7.95	10.78

--* statistics are missing because of failure of true solution on first step.

* statistics are a summary of all steps prior to a failure of the true solution.

complete. In the cases where the true solution failed on the first step, they are missing entirely (--*) and where the true solution has been initially successful, but has failed on other than the first step, the statistics are a summary of all steps up to the failure.

The various difficulties we have been discussing complicate the statistics and make it inappropriate to present a compact summary of the results, but the detailed statistics do allow us to make some definite observations and conclusions.

Of the five types of methods tested, only two, those based on backward differentiation formulas (DIFSUB, GEAR.REV3 and EPISODE) and the one based on second derivative multistep formulas (SDBASIC), are suitable as general purpose methods for stiff equations that arise in chemistry. The other three methods tested, although satisfactory on some problems, were very inefficient and/or unreliable on some of the test problems.

Of the three methods based on backward differentiation formulas, all performed very well on our tests, with GEAR.REV3 being the most efficient by a factor of about ten to twenty percent. One of the reasons for the improvement of GEAR.REV3 over DIFSUB in efficiency seems to be that it has a more relaxed local error control (as the larger local deceptions indicate) and, as a result, is slightly less reliable than DIFSUB. On the other hand EPISODE is less efficient on our test problems than GEAR.REV3, partly because of the larger number of LU decompositions and Jacobian evaluations required. This ranges up to more than twice as many on some problems. This behavior is probably caused by the difference in updating past values after a change of stepsize. Although the extra expense required by the strategy of EPISODE is probably justified for non-stiff equations, it can cause iteration difficulty for stiff problems and result in more updating of the iteration matrix (which requires a Jacobian evaluation followed by an LU decomposition). These extra LU decompositions would become more serious if larger systems of equations were being solved. EPISODE also experienced difficulty in the integration of two kinetics problems at the relaxed error tolerance when the problem became mathematically unstable.

Although the second derivative method SDBASIC is relatively efficient and reliable on all the test problems, it is not as efficient as those methods based on the backward differentiation formulas. The methods based on backward differentiation formulas seem to be consistently about three times more efficient than SDBASIC on the test problems. This can probably be attributed to the large values of $\left\|\frac{\partial^2 f}{\partial y^2}\right\|$ inherent in chemical kinetics problems, which can result in convergence difficulties in solving the implicit second derivative multistep formulas. (As with all methods tested, the iteration scheme is based on a modified Newton iteration, which, for second derivative formulas, assumes $\left\|h^2\frac{\partial^2 f}{\partial y^2}\right\|$ is small.) SDBASIC does require fewer steps to solve most of the problems than do the other methods tested and this generally leads to a more reliable control of global error (especially when an error-per-step control is used, as it is in our tests).

TRAPEX, which showed up quite well in our earlier comparisons, was unable to solve some of the kinetics problems. (For example CHM 9 and CHM 10 at 10^{-6}.) This can probably be attributed to the strong nonlinear coupling from smooth to transient components which is quite common in kinetics problems. This difficulty, also noted in our earlier comparisons, tends to result in very expensive integrations. For this reason TRAPEX is not recommended for the solution of kinetics equations, although methods based on the trapezoidal rule without extrapolation might be suitable for solving such problems at relaxed error tolerances.

The other methods tested, GENRK and IMPRK, were found to be unreliable and inefficient on some of the test problems. This confirms our earlier comparison results. Methods based on these techniques do not seem suitable for stiff equations arising in chemistry.

Although the results of these tests are consistent with our earlier comparisons, there are two difficulties not encountered in our current tests which can arise in stiff problems and cause some of the methods to fail. It is well known that methods based on backward differentiation formulas can be very inefficient on stiff problems which have eigenvalues of the Jacobian matrix close to the imaginary axis. In our earlier comparisons this was quite evident, and our results indicated that SDBASIC and TRAPEX both performed much better than DIFSUB on such problems. Of the test problems we have used, only CHM 1 and CHM 9 have non-real eigenvalues in parts of the integration. One must be aware of this difficulty and if trouble is encountered with a method based on backward differentiation formulas, then switching to another method, such as SDBASIC, should be considered.

A second difficulty that can arise is on problems with non-smooth solutions or rapidly changing characteristics. These types of problems can often lead to step control difficulties for DIFSUB, which can in turn result in failure. It was partly to overcome difficulties of this sort that GEAR.REV3 and EPISODE were developed.

In summary, we would like to make a specific recommendation to a practitioner who encounters a stiff chemical problem. He should first try a method based on backward differentiation formulas, such as GEAR.REV3, and if trouble arises, he should then switch to a second derivative method such as SDBASIC. This implies that a good subroutine library which is to allow for convenient solution of stiff chemical problems should contain a method based on backward differentiation formulas and another based on second derivative formulas.

REFERENCES

G. Bjurel, G. Dahlquist, B. Lindberg, S. Linde and L. Odén (1970) Survey of stiff ordinary differential equations, Report NA 70.11, Dept. of Information Processing, Royal Inst. of Tech., Stockholm.

J.C. Butcher (1964) Implicit Runge-Kutta processes, Math. Comp. 18, pp. 50-64.

G.D. Byrne and A.C. Hindmarsh (1975) A polyalgorithm for the numerical solution of ordinary differential equations, ACM Trans. on Math. Software 1, pp. 71-96.

G.J. Cooper (1969) The numerical solution of stiff differential equations, Computing Techniques in Biochemistry 2, Supplement, pp. S22-S29.

C.F. Curtiss and J.O. Hirschfelder (1952) Integration of Stiff Equations, Proc. Nat. Acad. Science, U.S., 38, pp. 235-243.

A.K. Datta (communicated by H.H. Robertson) (1967) An evaluation of the approximate inverse algorithm for numerical integration of stiff differential equations, Technical Report MSH/67/84, Imperial Chemical Industries Ltd., Cheshire.

G. Dahlquist (1963) A special stability problem for linear multistep methods, BIT 3, pp. 27-43.

W.H. Enright (1974) Second derivative multistep methods for stiff ordinary differential equations, SIAM J. Numer. Anal. 11, pp. 321-331.

W.H. Enright, R. Bedet, I. Farkas and T.E. Hull (1974) Test results on initial value methods for non-stiff ordinary differential equations, Dept. of Computer Science Tech. Rep. No. 68, University of Toronto, Toronto. (A summary of these results is to appear in the SIAM J. Numer. Anal. under the title "Test results on initial value methods for non-stiff ODE's" by W.H. Enright and T.E. Hull.)

W.H. Enright, T.E. Hull and B. Lindberg (1975) Comparing numerical methods for stiff systems of ODE's, BIT 15, pp. 10-49. (Also appeared as Dept. of Computer Science Tech. Rep. No. 69, University of Toronto.)

B.L. Ehle (1968) High order A-stable methods for the numerical solution of differential equations, BIT 8, pp. 276-278.

R.J. Field and R.M. Noyes (1974) Oscillations in chemical systems. IV. Limit cycle behavior in a model of a real chemical reaction, Journal of Chem. Phys. 60, pp. 1877-1884.

D. Garfinkel, D.W. Ching, M. Adelman and P. Clark (1966) Techniques and problems in the construction of computer models of biochemical systems including real enzymes, Annals New York Academy of Sciences, pp. 1054-1068.

C.W. Gear (1969) The automatic integration of stiff ordinary differential equations, Proceedings of IFIP Congress 1968, North Holland Publishing Company, Amsterdam, pp. 187-193.

C.W. Gear (1971a) Algorithm 407, DIFSUB for solution of ordinary differential equations, Comm. ACM 14, pp. 185-190.

C.W. Gear (1971b) Numerical initial value problems in ordinary differential equations, Prentice-Hall, Englewood Cliffs, N.J.

R.J. Gelinas (1972) Stiff systems of kinetics equations - a practitioner's view, J. Computational Phys. 9, pp. 222-236.

A.C. Hindmarsh (1974) GEAR: Ordinary differential equation system solver, UCID-3001, Rev. 3, Lawrence Livermore Laboratory, University of California, Livermore.

T.E. Hull, W.H. Enright, B.M. Fellen and A.E. Sedgwick (1972) Comparing numerical methods for ordinary differential equations, SIAM J. Numer. Anal. 9, pp. 603-637; errata 11, p. 681.

L. Lapidus, R.C. Aiken and Y.A. Liu (1974) The occurrence and numerical solution of physical and chemical systems having widely varying time constants, in Stiff Differential Systems (R.A. Willoughby ed.), Plenum Press, pp. 187-200.

J.D. Lawson and B.L. Ehle (1972) Improved generalized Runge-Kutta, Proceedings of Canadian Computer Conference, Session 72, pp. 223201-223213.

B. Lindberg (1971) On smoothing and extrapolation for the trapezoidal rule, BIT 11, pp. 29-52.

B. Lindberg (1972) IMPEX - a program package for solution of systems of stiff differential equations, Report NA72.50, Dept. of Information Processing, Royal Inst. of Tech., Stockholm.

W. Liniger and R.A. Willoughby (1967) Efficient numerical integration of stiff systems of ordinary differential equations, Technical Report RC-1970, Thomas J. Watson Research Center, Yorktown Heights, N.Y.

D. Luss and N.R. Amundson (1968) Stability of batch catalytic fluidized beds, AICHE Journal 14, pp. 211-221.

H.H. Robertson (1966) The solution of a set of reaction rate equations, in Numerical Analysis, An Introduction (J. Walsh ed.), Academic Press, London, pp. 178-182.

J.H. Seinfeld, L. Lapidus and M. Hwang (1970) Numerical integration of stiff ordinary differential equations, I and EC Fundamentals 9, pp. 266-275.

Appendix: Specification of Test Problems

CHM 1: $y_1'=-7.89\times10^{-10}y_1-1.1\times10^{7}y_1y_3$ $y_1(0)=1.76\times10^{-3}$

$y_2'=7.89\times10^{-10}y_1-1.13\times10^{9}y_2y_3$ $y_2(0) = 0$

$y_3'=7.89\times10^{-10}y_1-1.1\times10^{7}y_1y_3$

$+1.13\times10^{3}y_4 -1.13\times10^{9}y_2y_3$ $y_3(0)=0$

$y_4'=1.1\times10^{7}y_1y_3 -1.13\times10^{3}y_4$ $y_4(0)=0$

$x_f = 1000$ $h_0 = 5\times10^{-5}$

(chemical pyrolysis: Datta (1967))

CHM 2: $y_1'=-.04y_1+.01y_2y_3$ $y_1(0)=1$

$y_2'=400y_1-100y_2y_3-3000y_2^2$ $y_2(0)=0$

$y_3'=30y_2^2$ $y_3(0)=0$

$x_f = 40$ $h_0 = 10^{-5}$

(chemistry: Robertson (1966))

CHM 3: $y_1'=y_3-100y_1y_2$ $y_1(0)=1$

$y_2'=y_3+2y_4-100y_1y_2-2\times10^{4}y_2^2$ $y_2(0)=1$

$y_3'=-y_3+100y_1y_2$ $y_3(0)=0$

$y_4'=-y_4+10^{4}y_2^2$ $y_4(0)=0$

$x_f = 20$ $h_0 = 2.5\times10^{-5}$

(chemistry: Bjurel et al (1970))

CHM 4: $y_1'=-.013y_1-1000y_1y_3$ $y_1(0)=1$

$y_2'=-2500y_2y_3$ $y_2(0)=1$

$y_3'=-.013y_1-1000y_1y_3-2500y_2y_3$ $y_3(0)=0$

$x_f = 50$ $h_0 = 2.9\times10^{-4}$

(chemistry: Gear (1969))

CHM 5: $y_1'=.01-[1+(y_1+1000)(y_1+1)](.01+y_1+y_2)$ $y_1(0)=0$

$y_2'=.01-(1+y_2^2)(.01+y_1+y_2)$ $y_2(0)=0$

$x_f = 100$ $h = 10^{-4}$

(reactor kinetics: Liniger and Willoughby (1967))

CHM 6: $y_1'=1.3(y_3-y_1)+10400ky_2$ $y_1(0)=761$

$y_2'=1880[y_4-y_2(1+k)]$ $y_2(0)=0$

$y_3'=1752-269y_3+267y_1$ $y_3(0)=600$

$y_4'=.1+320y_2-321y_4$ $y_4(0)=0.1$

where $k = e^{(20.7-1500/y_1)}$

$x_f = 1000$ $h_0 = 10^{-4}$

(dynamics of a catalytic fluidized bed: Luss and Amundson (1968) and Lapidus et al (1974))

CHM 7: $y_1'=-y_1-y_1y_2+294y_2$ $y_1(0)=1$

$y_2'=y_1(1-y_2)/98-3y_2$ $y_2(0)=0$

$x_f = 240$ $h_0 = 10^{-2}$

(thermal decomposition of ozone: Lapidus et al (1974))

CHM 8: $y'=.2(y_2-y_1)$ $y_1(0)=0$

$y_2'=10y_1-(60-.125y_3)y_2+.125y_3$ $y_2(0)=0$

$y_3'=1$ $y_3(0)=0$

$x_f = 400$ $h_0 = 1.7\times10^{-2}$

(nuclear reactor theory: Liniger and Willoughby (1967))

CHM 9: $y_1'=77.27(y_2-y_1y_2+y_1-8.375\times10^{-6}y_1^2)$ $y_1(0)=4$

$y_2'=(-y_2-y_1y_2+y_3)/77.27$ $y_2(0)=1.1$

$y_3'=.161(y_1-y_3)$ $y_3(0)=4$

$x_f = 300$ $h_0= 10^{-3}$

(oscillating chemical system: Field and Noyes (1974))

CHM 10: $y_1' = 10^{11}(-3y_1y_2 + .0012y_4 - 9y_1y_3)$ $y_1(0) = 3.365\text{x}10^{-7}$

$y_2' = -3\text{x}10^{11}y_1y_2 + 2\text{x}10^{7}y_4$ $y_2(0) = 8.261\text{x}10^{-3}$

$y_3' = 10^{11}(-9y_1y_3 + .001y_4)$ $y_3(0) = 1.642\text{x}10^{-3}$

$y_4' = 10^{11}(3y_1y_2 - .0012y_4 + 9y_1y_3)$ $y_4(0) = 9.38\text{x}10^{-6}$

$x_f = 100$ $h_0 = 10^{-7}$

(enzyme kinetics: Garfinkel et al (1966))

On the Construction of Differential Systems for the Solution of Nonlinear Algebraic and Transcendental Systems of Equations

Dominic G. B. Edelen

1. INTRODUCTION

One of the more intriguing methods used in recent years for the numerical solution of nonlinear systems of equations $\underset{\sim}{y}(\underset{\sim}{x}) = \underset{\sim}{0}$ is the method of differential procedures. Such procedures replace the given equations $\underset{\sim}{y}(\underset{\sim}{x}) = \underset{\sim}{0}$ by a system of autonomous differential equations $d\underset{\sim}{x}(t)/dt = \underset{\sim}{f}(\underset{\sim}{x}(t))$ where $\underset{\sim}{f}(\underset{\sim}{x})$ is to be chosen so that "forward time-wise integration" of this differential system from some initial vector $\underset{\sim}{x}(0) = \underset{\sim}{a}$ will yield a vector $\underset{\sim}{x}(t)$ whose "large time limit" is a solution of the original system $\underset{\sim}{y}(\underset{\sim}{x}) = \underset{\sim}{0}$; that is $\underset{\sim}{y}(\lim_{\tau\to\infty} \underset{\sim}{x}(t)) = \underset{\sim}{0}$. Although such procedures are very useful, in view of the relative ease with which differential systems can be solved by high speed computers, there are three basic questions associated with their use. The first is that of establishing the initial data for which the solution of the differential system will have a finite large time limit; the ideal situation here would be that in which the large time limit exists for all finite initial data. The second is that of determining whether the large time limit points of the solutions of the differential system consist only of solutions of the original nonlinear system $\underset{\sim}{y}(\underset{\sim}{x}) = \underset{\sim}{0}$, or whether there are spurious limit points. And the third problem is that of determining how rapidly the solution of the differential system approaches its large time limit point.

Since the analysis given in this note takes a somewhat different approach to these problems than is reported in the current literature, it does not appear appropriate to cite specific references to the extensive body of work on this subject here. The readers familiar with these problems will, however, have little difficulty in perceiving parallels and implications with respect to the various methods in current use. We would be remiss, however, in not pointing out the theoretical significance of the method of associated differential systems in establishing existence and uniqueness of solutions of nonlinear elliptic systems of partial differential equations, as exemplified in the work of Eells and Sampson [1] and Hartman [2] on the problem of harmonic maps of Riemannian manifolds.

We first give a general procedure for constructing families of differential systems for which every solution with finite initial data has the zero vector as its large time limit point. This is achieved by means of Liapunov's method for determining asymptotic stability. Application of this procedure to the problem of solving $\underset{\sim}{y}(\underset{\sim}{x}) = \underset{\sim}{0}$ is shown to yield a vector $\underset{\sim}{x}(t)$ for any finite initial data such that the large time limit of $\underset{\sim}{x}(t)$ satisfies $\underset{\sim}{y}(\underset{\sim}{x}) = \underset{\sim}{0}$ provided $\Delta(\underset{\sim}{x}) = \det(\partial y_i/\partial x_j) \neq 0$ for all finite $\underset{\sim}{x}$. If $\Delta(\underset{\sim}{x}) = 0$ on some non-empty set D of points, then the same procedure as used in the case $\Delta(\underset{\sim}{x}) \neq 0$ is shown to imply that the norm of $d\underset{\sim}{x}/dt$ can grow without bound and that the norm of $\underset{\sim}{x}(t)$ can grow without bound if $\| \underset{\sim}{y}(\underset{\sim}{x}) \|$ does not tend to infinity with $\| \underset{\sim}{x} \|$. We construct two different classes of procedures that generate solutions of $\underset{\sim}{y}(\underset{\sim}{x}) = \underset{\sim}{0}$ when $\Delta(\underset{\sim}{x}) = 0$ and $\| \underset{\sim}{y}(\underset{\sim}{x}) \| \to \infty$ as $\| \underset{\sim}{x} \| \to \infty$. Additional conditions on the initial values have to be imposed if $\| \underset{\sim}{y}(\underset{\sim}{x}) \| \not\to \infty$ as $\| \underset{\sim}{x} \| \to \infty$. Unfortunately, some of the large time limit points of the solutions of these differential systems belong to the set $\Delta(\underset{\sim}{x}) = 0$, so that the method yields spurious limit points as well as solutions of $\underset{\sim}{y}(\underset{\sim}{x}) = \underset{\sim}{0}$. We show that the procedures can be adjusted so as to obtain exponential approach to the large time limit points, and, of more importance, we can also obtain exponential approach in finite time. Specifically, we give methods for which $\| \underset{\sim}{y}(\underset{\sim}{x}(t)) \|^2 \leq \lambda\ 2\ \exp(-\tan(kt+c))$ where k is a positive constant and c is a constant that is determined by the initial data.

2. A GLOBAL ASYMPTOTIC STABILITY THEOREM

Let $\underset{\sim}{x} \equiv (x_1,\ldots,x_N)$ be a generic element of an N-dimensional vector space E_N with inner product $(\underset{\sim}{x},\underset{\sim}{y})$ and norm $\| \underset{\sim}{x} \| = (\underset{\sim}{x},\underset{\sim}{x})^{\frac{1}{2}}$. The basis of the methods given in this note is the following global asymptotic stability theorem.

Every solution of the system of autonomous differential equations

$$\frac{d}{dt}\underset{\sim}{y}(t) = -\underset{\sim}{F}(\underset{\sim}{x}(t)) \tag{2.1}$$

with finite initial data is such that $\lim_{t\to\infty} \underset{\sim}{y}(t) = \underset{\sim}{0}$ *provided* $\underset{\sim}{F}(\underset{\sim}{y})$ *is a* C^1 *vector-valued function that is given by*

$$\underset{\sim}{F}(\underset{\sim}{y}) = \underset{\sim}{\nabla}_W \phi(\underset{\sim}{W}(\underset{\sim}{y})) + \underset{\sim}{U}(\underset{\sim}{W}(\underset{\sim}{y})) \tag{2.2}$$

for some C^1 *vector-valued function* $\underset{\sim}{U}(\underset{\sim}{W})$ *such that*

$$\underset{\sim}{U}(\underset{\sim}{0}) = \underset{\sim}{0}\ ,\quad (\underset{\sim}{W},\underset{\sim}{U}(\underset{\sim}{W})) = 0 \tag{2.3}$$

and some C^2 *scalar-valued function*

$$\phi(\underset{\sim}{W}) = \int_0^1 \{P(\lambda\underset{\sim}{W}) - K(\lambda\underset{\sim}{W})\}\frac{d\lambda}{\lambda} \tag{2.4}$$

where (i)

$$\underset{\sim}{W}(\underset{\sim}{y}) = \underset{\sim}{\nabla}_y V(\underset{\sim}{y})\ , \tag{2.5}$$

(ii) $V(\underset{\sim}{y})$ *is a* C^3 *scalar-valued function such that*

$$V(\underset{\sim}{y}) \geq 0 \ , \ V(\underset{\sim}{y}) \to \infty \text{ as } \|\underset{\sim}{y}\| \to \infty \tag{2.6}$$

$$V(\underset{\sim}{y}) = 0 \iff \underset{\sim}{y} = \underset{\sim}{0} \ , \tag{2.7}$$

$$\underset{\sim y}{\nabla} V(\underset{\sim}{y}) = 0 \iff \underset{\sim}{y} = \underset{\sim}{0} \ , \tag{2.8}$$

and (iii) $P(\underset{\sim}{W})$ *and* $K(\underset{\sim}{W})$ *are* C^2 *scalar-valued functions such that*

$$K(\underset{\sim}{W}) \leq 0 \tag{2.9}$$

$$K(\underset{\sim}{W}) = 0 \iff \underset{\sim}{W} = \underset{\sim}{0} \ , \tag{2.10}$$

$$P(\underset{\sim}{W}) \geq 0 \ , \tag{2.11}$$

$$P(\underset{\sim}{0}) = 0 \ . \tag{2.12}$$

Proof. We first note that (2.9) - (2.12) imply

$$\underset{\sim W}{\nabla}(P-K)\Big|_{\underset{\sim}{W}=\underset{\sim}{0}} = \underset{\sim}{0} \ , \tag{2.13}$$

while (2.6) - (2.8) are certainly satisfied for $V(\underset{\sim}{y}) = \frac{1}{2}(\underset{\sim}{y},\underset{\sim}{y})$. Since $V(\underset{\sim}{y})\varepsilon C^3$, we have $W(\underset{\sim}{y}) = \underset{\sim y}{\nabla} V(\underset{\sim}{y})\varepsilon C^2$, and hence (2.2), (2.4) and the continuity conditions satisfied by $\underset{\sim}{U}(\underset{\sim}{W})$, $P(\underset{\sim}{W})$ and $K(\underset{\sim}{W})$ show that $\underset{\sim}{F}(\underset{\sim}{y})\varepsilon C^1$. It also follows from (2.2) - (2.5), (2.8) and (2.13) that $\underset{\sim}{F}(\underset{\sim}{y}) = \underset{\sim}{0} \iff \underset{\sim}{W}(\underset{\sim}{y}) = 0 \iff \underset{\sim}{y} = \underset{\sim}{0}$, and hence $\underset{\sim}{y} = \underset{\sim}{0}$ is the only critical point of the system (2.1). A straightforward calculation and use of (2.1) gives us

$$\frac{d}{dt} V(\underset{\sim}{y}(t)) = (\underset{\sim y}{\nabla} V(\underset{\sim}{y}), d\underset{\sim}{y}(t)/dt) = -(\underset{\sim y}{\nabla} V, \underset{\sim}{F}) \ , \tag{2.14}$$

and hence (2.6) - (2.8) show that $V(\underset{\sim}{y})$ will be a global Liapunov function for the system (2.1) that implies the global asymptotic stability of $\underset{\sim}{y} = \underset{\sim}{0}$ provided [3]

$$\psi(\underset{\sim}{y}) = (\underset{\sim y}{\nabla} V(\underset{\sim}{y}), \underset{\sim}{F}(\underset{\sim}{y})) \tag{2.15}$$

satisfies the conditions

$$\psi(\underset{\sim}{y}) \geq 0 \ , \tag{2.16}$$

$$\psi(\underset{\sim}{y}) = 0 \iff \underset{\sim}{y} = \underset{\sim}{0} \ . \tag{2.17}$$

Now, (2.5) and (2.8) imply that $\underset{\sim}{W}(\underset{\sim}{y}) = \underset{\sim}{0} \Leftrightarrow \underset{\sim}{y} = \underset{\sim}{0}$, while (2.2) shows that we actually have $\underset{\sim}{F}(\underset{\sim}{y}) = \underset{\sim}{F}^*(\underset{\sim}{W}(\underset{\sim}{y}))$, where $\underset{\sim}{F}^*(\underset{\sim}{W})$ is of class C^1 in $\underset{\sim}{W}$. Accordingly (2.15) - (2.17) can be rewritten in the equivalent forms

$$\psi(\underset{\sim}{y}) = \psi^*(\underset{\sim}{W}(\underset{\sim}{y})) = (\underset{\sim}{W}(\underset{\sim}{y}), F^*(\underset{\sim}{W}(\underset{\sim}{y}))) , \qquad (2.18)$$

$$\psi^*(\underset{\sim}{W}) \geq 0 , \qquad (2.19)$$

$$\psi^*(\underset{\sim}{W}) = 0 \Leftrightarrow \underset{\sim}{W} = \underset{\sim}{0} . \qquad (2.20)$$

We have shown in a previous paper [4] that the general solution of the inequality

$$(\underset{\sim}{W}, \underset{\sim}{F}^*(\underset{\sim}{W})) + K(\underset{\sim}{W}) \geq 0 , \qquad (2.21)$$

for $\underset{\sim}{F}^*(\underset{\sim}{W}) \varepsilon\ C^1$ and $K(\underset{\sim}{W}) \varepsilon C^2$, is given by

$$\underset{\sim}{F}^*(\underset{\sim}{W}) = \underset{\sim}{\nabla}_W \phi(\underset{\sim}{W}) + \underset{\sim}{U}(\underset{\sim}{W}) , \qquad (2.22)$$

where (2.3), (2.4), (2.11), $P(\underset{\sim}{0}) = K(\underset{\sim}{0}) = 0$ and the given continuity conditions on $\underset{\sim}{U}(\underset{\sim}{W})$, $P(\underset{\sim}{W})$ and $K(\underset{\sim}{W})$ are satisfied. In fact, the relation between $\underset{\sim}{F}^*$, K and P is

$$(\underset{\sim}{W}, \underset{\sim}{F}^*(\underset{\sim}{W})) + K(\underset{\sim}{W}) = P(\underset{\sim}{W}) \geq 0 , \qquad (2.23)$$

and hence equality holds in (2.21) if and only if $P(\underset{\sim}{W}) \equiv 0$. It thus follows, on combining (2.18) with (2.21) and (2.23), that

$$\psi^*(\underset{\sim}{W}) \geq - K(\underset{\sim}{W}) , \qquad (2.24)$$

with equality holding if and only if $P(\underset{\sim}{W}) \equiv 0$. The inequalities (2.19) and (2.20) are thus satisfied provided $K(\underset{\sim}{W})$ is such that (2.9) and (2.10) hold. We must also have satisfaction of (2.12) since $P(\underset{\sim}{0}) = K(\underset{\sim}{0}) = 0$. The function $V(\underset{\sim}{y})$ is thus established as a Liapunov function for the system (2.1) with $dV(\underset{\sim}{y}(t))/dt = 0$ if and only if $\underset{\sim}{y} = \underset{\sim}{0}$, and the theorem is established.

It is of interest to note that the above theorem characterizes the most general system of autonomous differential equations (2.1) for which $\underset{\sim}{F}(\underset{\sim}{y})$ is of class C^1 and $\underset{\sim}{y} = \underset{\sim}{0}$ is a globally asymptotically stable critical point with a global Liapunov function $V(\underset{\sim}{y})$ of class C^3 .

3. DIFFERENTIAL METHODS LIMITED TO INVERTIBLE SYSTEMS

As before, let $\underset{\sim}{x}$ be a generic element of E_N and let $\underset{\sim}{y}(\underset{\sim}{x})$ be a C^1 mapping of E_N into E_N . Further, let $\underset{\sim}{x}(t)$ be a C^1 mapping of $[0,\infty)$ into E_N , then $\underset{\sim}{y}^*(t) = \underset{\sim}{y}(\underset{\sim}{x}(t))$ is a C^1 mapping of $[0,\infty)$ into E_N . With $\underset{\sim}{F}(\underset{\sim}{y}^*(t))$ chosen in accordance with the hypotheses of the previous theorem, the system of autonomous differential equations

$$\frac{d}{dt}\underset{\sim}{y}^*(t) = -\underset{\sim}{F}(\underset{\sim}{y}^*(t)) \tag{3.1}$$

has a solution $\underset{\sim}{y}^*(t)$, by forward stepping procedures for all finite initial data

$$\underset{\sim}{y}^*_o = \underset{\sim}{y}^*(0) , \tag{3.2}$$

which tends to the zero vector for sufficiently large t . What we want, however, is a system of equations for the determination of $\underset{\sim}{x}(t)$, rather than $\underset{\sim}{y}^*(t) = \underset{\sim}{y}(\underset{\sim}{x}(t))$, which tends to solutions of

$$\underset{\sim}{y}(\underset{\sim}{x}) = \underset{\sim}{0} \tag{3.3}$$

for sufficiently large t . A formal substitution of $\underset{\sim}{y}^*(t) = \underset{\sim}{y}(\underset{\sim}{x}(t))$ into (3.1) yields

$$\sum_{j=1}^{N} \frac{\partial y_i}{\partial x_j}\frac{d}{dt}x_j(t) = -F_i(\underset{\sim}{y}(\underset{\sim}{t}))) \equiv -F_i^*(\underset{\sim}{x}(t)) , \tag{3.4}$$

from which it is clear that a system of differential equations can be obtained for $\underset{\sim}{x}(t)$. The large time limit properties of the solutions of (3.4) are, however, quite different from those of solutions of (3.1). In fact, there are two distinct cases that arise.

The first case is delineated by satisfaction of the condition

$$\Delta(\underset{\sim}{x}) \overset{\text{def}}{=} \det(\partial y_i/\partial x_j) \neq 0 \tag{3.5}$$

for every $\underset{\sim}{x} \varepsilon E_N$, and that $\underset{\sim}{y}(\underset{\sim}{x})$ is of class C^2 . In this case, $\underset{\sim}{y} = \underset{\sim}{y}(\underset{\sim}{x})$ is a one-to-one, onto map of E_N into itself (globally invertible). It thus follows that $V(\underset{\sim}{y}(\underset{\sim}{x})) \to \infty$ as $\| \underset{\sim}{x} \| \to \infty$ if $V(\underset{\sim}{y}) \to \infty$ as $\| \underset{\sim}{y} \| \to \infty$. Under satisfaction of (3.5), the system (3.4) can be written in the equivalent form

$$\frac{d}{dt} x_i(t) = \sum_{j=1}^{N} Y_{ij}(\underset{\sim}{x}) F_j^*(x) , \qquad (3.6)$$

where $Y_{ij}(\underset{\sim}{x})$ are the entries of the inverse of the matrix $((\partial y_i/\partial x_j))$. Since $\underset{\sim}{y}(\underset{\sim}{x}) \varepsilon C^2$, the functions $Y_{ij}(\underset{\sim}{x})$ belong to C^1 so that the right-hand sides of (3.6) are C^1 functions of $\underset{\sim}{x}$.

Let $V(\underset{\sim}{y})$ be the Liapunov function for the system (3.1) that is used in the construction of the vector-valued function $\underset{\sim}{F}(\underset{\sim}{y})$. We then have

$$V^*(\underset{\sim}{x}) = V(\underset{\sim}{y}(\underset{\sim}{x})) \geq 0 , \; V^*(\underset{\sim}{x}) \to \infty \text{ as } \| \underset{\sim}{x} \| \to \infty , \qquad (3.7)$$

$$V^*(\underset{\sim}{x}) = 0 \iff \underset{\sim}{y}(\underset{\sim}{x}) = 0 , \qquad (3.8)$$

$$\frac{\partial V^*}{\partial x_i} = \sum_j \frac{\partial V}{\partial y_j} \frac{\partial y_j}{\partial x_i} = 0 \iff \frac{\partial V}{\partial y_j} = 0 \iff \underset{\sim}{y}(\underset{\sim}{x}) = 0 \qquad (3.9)$$

from (2.6) - (2.8), and

$$\frac{dV^*}{dt} = (\underset{\sim x}{\nabla} V^*, \frac{d}{dt} \underset{\sim}{x}) = (\underset{\sim y}{\nabla} V, \frac{d}{dt} \underset{\sim}{y}) \leq K(\underset{\sim}{y}(\underset{\sim}{x})) = K^*(\underset{\sim}{x}) \leq 0 , \qquad (3.10)$$

$$K^*(\underset{\sim}{x}) = 0 \iff \underset{\sim}{y}(\underset{\sim}{x}) = 0 \qquad (3.11)$$

from (2.10), (2.14), (2.15) and (2.24). Thus $V^*(\underset{\sim}{x})$ is a global Liapunov function for the system (3.6) whose critical points are those $\underset{\sim}{x}$ for which $\underset{\sim}{y}(\underset{\sim}{x}) = \underset{\sim}{0}$. Convergence of all solutions of (3.6) to solutions of $\underset{\sim}{y}(\underset{\sim}{x}) = \underset{\sim}{0}$, for sufficiently large t thus follows for all finite initial data $\underset{\sim}{x}_o = \underset{\sim}{x}(0)$. In fact, it follows that the solution of $\underset{\sim}{y}(\underset{\sim}{x}) = \underset{\sim}{0}$ is unique in E_n for this case. Unfortunately, the utility of the differential system (3.6) is not overly great since its use requires that we test to make sure that $\Delta(\underset{\sim}{x}) \neq 0$, and this is often of equal or greater difficulty than just solving $\underset{\sim}{y}(\underset{\sim}{x}) = \underset{\sim}{0}$. Clearly, what is required is a method whose convergence is not tied to an *a priori* test such as $\Delta(\underset{\sim}{x}) \neq 0$. This indeed is a real, rather than a spurious concern in view of the results we establish below for the second case where $\Delta(\underset{\sim}{x}) = 0$ at one or more points in E_N.

The second case is characterized by the fact that $\Delta(\underset{\sim}{x}) = 0$ at one or more points in E_N. Analysis of this case is most easily pursued by introducing the following point sets:

$$\mathcal{D} = \{\underset{\sim}{x}\varepsilon E_N | \det(\partial y_i/\partial x_j) = 0\} , \tag{3.12}$$

$$\mathcal{J} = \{\underset{\sim}{x}\varepsilon E_N | \quad \|\nabla_{\underset{\sim}{x}} V^*\| = 0 , \quad \|\underset{\sim}{y}(\underset{\sim}{x})\| \neq 0\} , \tag{3.13}$$

$$\chi = \{\underset{\sim}{x}\varepsilon E_N | \underset{\sim}{y}(\underset{\sim}{x}) = \underset{\sim}{0}\} , \tag{3.14}$$

where $V^*(\underset{\sim}{x}) = V(\underset{\sim}{y}(\underset{\sim}{x}))$ and $V(\underset{\sim}{y})$ is the Liapunov function for the system (3.1) that is used in the construction of $F(\underset{\sim}{y})$. Since

$$\frac{\partial V^*}{\partial x_i} = \sum_{j=1}^{N} \frac{\partial V}{\partial y_j} \frac{\partial y_j}{\partial x_i}$$

and $\nabla_{\underset{\sim}{y}} V = \underset{\sim}{0} \iff \underset{\sim}{y}(\underset{\sim}{x}) = 0 \iff \underset{\sim}{x}\varepsilon\chi$, we see that $\underset{\sim}{x}\varepsilon\mathcal{J}$ only if $\underset{\sim}{x}\varepsilon\mathcal{D}$; that is $\mathcal{J} \subset \mathcal{D}$. It also follows that

$$\frac{d}{dt} V^* = (\nabla_{\underset{\sim}{x}} V^*, \frac{d}{dt} \underset{\sim}{x}) = (\nabla_{\underset{\sim}{y}} V, \frac{d}{dt} \underset{\sim}{y}) \leq K(\underset{\sim}{y}(\underset{\sim}{x})) = K^*(\underset{\sim}{x}) \leq 0, \tag{3.15}$$

$$K(\underset{\sim}{y}(\underset{\sim}{x})) = K(\nabla_{\underset{\sim}{y}} V(\underset{\sim}{y}(\underset{\sim}{x})) = K^*(\underset{\sim}{x}) = 0 \iff \underset{\sim}{x}\varepsilon\chi . \tag{3.16}$$

However, for each $\underset{\sim}{x}\varepsilon\mathcal{J}$, $\|\nabla_{\underset{\sim}{x}} V^*\| = 0$. Since $\chi \cap \mathcal{J}$ is empty, it follows from (3.15) that $\|\frac{d}{dt} \underset{\sim}{x}\|$ is unbounded for every $\underset{\sim}{x}\varepsilon\mathcal{J}$. An integration of the system (3.4) can thus lead to unbounded values for dx_i/dt and global asymptotic stability of the system (3.4) can fail to obtain. Further, it is not necessarily true that $V(\underset{\sim}{y}) \to \infty$ as $\|\underset{\sim}{y}\| \to \infty$ implies that $V^*(\underset{\sim}{x}) = V(\underset{\sim}{y}(\underset{\sim}{x})) \to \infty$ as $\|\underset{\sim}{x}\| \to \infty$, for we lose the one-to-one character of $\underset{\sim}{y} = \underset{\sim}{y}(\underset{\sim}{x})$. However, (3.15), (3.16) show that $V^*(\underset{\sim}{x})$ is either constant or decreasing on any solution of the system (3.5), and hence $\underset{\sim}{x}(t)$ can go to infinity for certain initial data if $V^*(\underset{\sim}{x})$ decreases as $\| \underset{\sim}{x} \| \to \infty$ in any given direction. Thus, existence of finite large time limit points can be lost. In addition, continuity of $d\underset{\sim}{x}/dt$ as functions of $\underset{\sim}{x}$ is lost as $\underset{\sim}{x}(t)$ approaches the point set $\mathcal{D}$, and hence solutions of (3.4) may not be defined for large t. For example $y(x) = x^2 + 1$, $V(y) = \frac{1}{2} y^2$, $dV/dy = W = y$, $\phi(W) = \frac{k}{2} W^2 = k\, V(y)$, so that (3.4) becomes $2x \frac{dx}{dt} = - k(x^2+1)$ with solutions $x^2 + 1 = (x_o^2+1)\exp(-kt)$.

It should now be clear that the differential procedures based on (3.4) do not have general validity except in the case where it is established that $\Delta(\underset{\sim}{x}) \neq 0$ for all $\underset{\sim}{x}$. On the other hand, procedures for testing whether $\Delta(\underset{\sim}{x}) \neq 0$ for all $\underset{\sim}{x}$ can be even more difficult than the original problem of solving $\underset{\sim}{y}(\underset{\sim}{x}) = \underset{\sim}{0}$. If it be easily determined that $\Delta(\underset{\sim}{x}) \neq 0$ for all $\underset{\sim}{x}$, then the procedures based on (3.4) can be used, but not all problems present this pleasant circumstance. The next Section develops a general class of procedures that are applicable to all problems for which $V^*(\underset{\sim}{x}) = V(\underset{\sim}{y}(\underset{\sim}{x})) \to \infty$ as $\| \underset{\sim}{x} \| \to \infty$.

The above analysis indicates that differential procedures based on (3.4) run into difficulty for the same reason that a Newton iteration procedure runs into difficulty; namely, because the inverse of $((\partial y_i/\partial x_j))$ becomes unbounded if $\Delta(\underset{\sim}{x}) = 0$. It is also clear that the Davidenko-Branin [2] method suffers from the same difficulty, (we may take $V(\underset{\sim}{y}) = \frac{1}{2} (\underset{\sim}{y},\underset{\sim}{y})$, $\underset{\sim}{W} = \underset{\sim}{\nabla}_y V = \underset{\sim}{y}$, $\phi(\underset{\sim}{W}) = \frac{k}{2} (\underset{\sim}{W},\underset{\sim}{W}) = k\, V(\underset{\sim}{y})$ so that $\underset{\sim}{F}(\underset{\sim}{y}) = k\, \underset{\sim}{y}$ and (3.4) becomes

$$\sum_{j=1}^{N} \frac{\partial y_i}{\partial x_j} \frac{dx_j}{dt} = - k y_i) .$$

4. DIFFERENTIAL METHODS FOR NON-INVERTIBLE SYSTEMS

It was shown in the last Section that there are many

differential procedures for solving $y(x) = 0$ when $\Delta(x) \neq 0$ throughout E_N. We give two alternative collections of differential procedures in this Section for solving $y(x) = 0$ that work for $\Delta(x) = 0$ on a non-empty set $\mathcal{D}$ in E_N, provided $\| x \| \to \infty$ implies $\| y(x) \| \to \infty$. Alternatives to these growth conditions are also given. Unfortunately, there is a price to be paid for universality, and in these instances it comes about as spurious limit points.

Let $V(y)$ be a C^2 scalar-valued function such that

$$V(y) \geq 0 \ , \ V(y) = 0 \Longleftrightarrow y = 0 \ , \tag{4.1}$$

$$\nabla_y V(y) = 0 \Longleftrightarrow y = 0 \ , \ V(y) \to \infty \ \text{as} \ \| y \| \to \infty \ , \tag{4.2}$$

and define the function $V^*(x)$ by

$$V^*(x) = V(y(x)) \tag{4.3}$$

for $y(x)$ a given C^2 vector-valued function. Since $\| y(x) \| \to \infty$ as $\| x \| \to \infty$, we have $V^*(x) = V(y(x) \to \infty$ as $\| x \| \to \infty$. Let

$$F^*(x) = F(W(x)) \tag{4.4}$$

be defined by (2.2), *where* $W(x)$ *is now given by*

$$W(x) = \nabla_x V^*(x) \ , \tag{4.5}$$

and (2.3), (2.4), (2.9) - (2.13) hold. We consider the system of autonomous differential equations

$$\frac{d}{dt} x(t) = - F^*(x) \ . \tag{4.6}$$

Use of this system shows that

$$\frac{d}{dt} V^*(x) = - (\nabla_x V^*, F^*(x)) = - (W, F(W)) \ . \tag{4.7}$$

The same reasoning as that given in Section 2 now yields

$$\frac{d}{dt} V^*(x) \leq K(W) \leq 0 \tag{4.8}$$

with

$$K(\underset{\sim}{W}) = 0 \iff \underset{\sim}{W} = \underset{\sim}{0} . \tag{4.9}$$

It also follows from (4.4) and the results established in Section 2 that

$$\underset{\sim}{F}^*(\underset{\sim}{x}) = \underset{\sim}{F}(\underset{\sim}{W}(\underset{\sim}{x})) = \underset{\sim}{0} \iff \underset{\sim}{W} = \underset{\sim}{0} , \tag{4.10}$$

and hence dV^*/dt vanishes only at the critical points of the system (4.6) which are the points $\underset{\sim}{W}(\underset{\sim}{x}) = \underset{\sim}{0}$. However, we saw in the last Section that $\underset{\sim}{W}(\underset{\sim}{x}) = \nabla_{\underset{\sim}{x}} V^*(\underset{\sim}{x}) = \underset{\sim}{0}$ holds for all $x \varepsilon \chi \cup J$. Accordingly, the same reasoning as used in Section 3 shows that every solution of the system (4.6) with initial data in the complement of J will have a large time limit point in $\chi \cup M$. Here M is the subset of joints of J for which $V^*(\underset{\sim}{x})$ has a relative minimum. It also follows from the fact that the set χ is the set of absolute minima of $V^*(\underset{\sim}{x})$, that every isolated point $\bar{\underset{\sim}{x}}$ of χ belongs to an open neighborhood $N(\bar{\underset{\sim}{x}})$ of initial data which contains $\bar{\underset{\sim}{x}}$ and is such that every solution of (4.6) with initial data in $N(\bar{\underset{\sim}{x}})$ will have $\bar{\underset{\sim}{x}}$ as its large time limit point. If χ contains an arcwise connected subset A , then there exists an arcwise connected open set $N(A)$ which contains A as a proper subset such that each solution of (4.6) with initial data in $N(A)$ will tend, for large t , to a point on the boundary of A if the initial point does not belong to A , and will remain fixed in A if the initial data point belongs to A .

As examples, take $V(\underset{\sim}{y}) = \frac{1}{2}(\underset{\sim}{y},\underset{\sim}{y})$, so that $V^*(\underset{\sim}{x}) = \frac{1}{2}(\underset{\sim}{y}(\underset{\sim}{x}),\underset{\sim}{y}(\underset{\sim}{x}))$; that is $V^*(x)$ is proportional to the square of the error in solving $\underset{\sim}{y}(\underset{\sim}{x}) = \underset{\sim}{0}$. We then have $\underset{\sim}{W}(\underset{\sim}{x}) = \frac{1}{2}\nabla_{\underset{\sim}{x}}(\underset{\sim}{y}(\underset{\sim}{x}),\underset{\sim}{y},(\underset{\sim}{x}))$ so that $\underset{\sim}{W}(\underset{\sim}{x})$ is a vector that points in the direction of maximal increase in the error. With $\underset{\sim}{F}(\underset{\sim}{W}) = k\underset{\sim}{W}$, the systems (4.6) and (4.7) become

$$\frac{d}{dt}\underset{\sim}{x}(t) = -k\underset{\sim}{W}(\underset{\sim}{x}) = -\frac{k}{2}\nabla_{\underset{\sim}{x}}(\underset{\sim}{y}(\underset{\sim}{x}),\underset{\sim}{y}(\underset{\sim}{x}))$$

$$\frac{d}{dt}\frac{1}{2}(\underset{\sim}{y}(\underset{\sim}{x}),\underset{\sim}{y}(\underset{\sim}{x})) = -k(\underset{\sim}{W}(\underset{\sim}{x}),\underset{\sim}{W}(\underset{\sim}{x})) .$$

For these choices of $V(\underset{\sim}{y})$ and $\underset{\sim}{F}(\underset{\sim}{W})$, the differential system (4.6) shows that $\underset{\sim}{x}(t)$ evolves with time along an orbit of E_N whose tangent is always in the direction of maximal decrease of the square of the error, $\frac{1}{2}(\underset{\sim}{y}(\underset{\sim}{x}),\underset{\sim}{y}(\underset{\sim}{x}))$.

Although these results provide methods that give solutions of $\underset{\sim}{y}(\underset{\sim}{x}) = \underset{\sim}{0}$ whenever $\|\underset{\sim}{x}\| \to \infty$ implies $\|\underset{\sim}{y}(\underset{\sim}{x})\| \to \infty$, (i.e. without knowledge that $\Delta(\underset{\sim}{x}) \neq 0$ for all $\underset{\sim}{x}$) an *a priori* elimination of spurious limit points requires that we characterize the set of joints J. Of course, the spurious limit points need not be known beforehand since it is easy enough to test whether a large time limit point of the system (4.6) satisfies $\underset{\sim}{y}(\underset{\sim}{x}) = \underset{\sim}{0}$ once the limit point is known. On the other hand, characterization of J requires that we solve $\underset{\sim}{\nabla}_x V^*(\underset{\sim}{x}) = \underset{\sim}{0}$, which can be as bad as solving $\underset{\sim}{y}(\underset{\sim}{x}) = \underset{\sim}{0}$. We therefore present an alternative collection of methods that is not dependent on knowledge of J for elimination of spurious limit points.

Let $\underset{\sim}{y}(\underset{\sim}{x})$ be a given C^1 vector-valued function and let

$$\underset{\approx}{J}(\underset{\sim}{x}) = ((\partial y_i(\underset{\sim}{x})/\partial x_j)) \tag{4.11}$$

denote the Jacobian matrix of $\underset{\sim}{y}(\underset{\sim}{x})$ with respect to $\underset{\sim}{x}$. For the remainder of this Section we view $\underset{\sim}{x}$ as a column matrix with N entries and $\underset{\sim}{x}^T$, the transpose of $\underset{\sim}{x}$, as a row matrix. Further, let $\underset{\approx}{C}(\underset{\sim}{x})$ denote the matrix of cofactors of the matrix $\underset{\approx}{J}(\underset{\sim}{x})$, so that

$$\underset{\approx}{J}(\underset{\sim}{x})\ \underset{\approx}{C}(\underset{\sim}{x})^T = \Delta(\underset{\sim}{x})\ \underset{\approx}{E}\ , \tag{4.12}$$

where $\underset{\approx}{E}$ denotes the N-by-N identity matrix and

$$\Delta(\underset{\sim}{x}) = \det(\underset{\approx}{J}(\underset{\sim}{x})) = \det(\partial y_i/\partial x_j)\ . \tag{4.13}$$

Choose $\underset{\sim}{F}(\underset{\sim}{y})$ in accordance with the asymptotic stability theorem given in Section 2 in terms of the functions $U(\underset{\sim}{W})$, $V(\underset{\sim}{y})$, $P(\underset{\sim}{W})$, $K(\underset{\sim}{W})$, where we now have $\underset{\sim}{W} = \underset{\sim}{\nabla}_y V(\underset{\sim}{y})$. A convenient system of autonomous differential equations to be used under these conditions is

$$\frac{d}{dt} \underset{\sim}{x}(t) = - \Delta(\underset{\sim}{x}) \; \underset{\approx}{C}(\underset{\sim}{x})^T \; \underset{\sim}{F}^*(\underset{\sim}{x}) \tag{4.14}$$

where $\underset{\sim}{F}^*(\underset{\sim}{x}) = \underset{\sim}{F}(\underset{\sim}{y}(\underset{\sim}{x}))$. We then have

$$\begin{aligned} \frac{d}{dt} \underset{\sim}{y} = \underset{\approx}{J}(\underset{\sim}{x}) \frac{d}{dt} \underset{\sim}{x} &= - \Delta(\underset{\sim}{x}) \; \underset{\approx}{J}(\underset{\sim}{x}) \; \underset{\approx}{C}(\underset{\sim}{x})^T \; \underset{\sim}{F}^*(\underset{\sim}{x}) \\ &= - (\Delta(\underset{\sim}{x}))^2 \; \underset{\sim}{F}(\underset{\sim}{y}) \; , \end{aligned} \tag{4.15}$$

when (4.12) is used. The only difference between (4.15) and (2.1) is the factor $(\Delta(\underset{\sim}{x}))^2$ on the right-hand side of (4.14). Hence the critical points of the system (4.15) consist of those for which $\underset{\sim}{y} = \underset{\sim}{y}(\underset{\sim}{x}) = \underset{\sim}{0}$ and, in addition, those for which $\Delta(\underset{\sim}{x}) = 0$; that is, the set of points $\chi \cup \mathcal{D}$. Likewise, since the matrix $\underset{\approx}{C}(\underset{\sim}{x})$ is singular only if $\Delta(\underset{\sim}{x}) = 0$, the system (4.14) possesses exactly the same critical points, namely $\chi \cup \mathcal{D}$. A straightforward calculation and the results established in Section 2 give

$$\begin{aligned} \frac{d}{dt} V(\underset{\sim}{y}) &= (\nabla_{\underset{\sim}{y}} V)^T \frac{d}{dt} \underset{\sim}{y} \\ &= - (\Delta(\underset{\sim}{x}))^2 (\nabla_{\underset{\sim}{y}} V, \underset{\sim}{F}) \le (\Delta(\underset{\sim}{x}))^2 \; K(\nabla_{\underset{\sim}{y}} V) \le 0 \end{aligned} \tag{4.16}$$

$$\frac{d}{dt} V(\underset{\sim}{y}) = 0 \Leftrightarrow \underset{\sim}{y} = \underset{\sim}{0} \quad \text{or} \quad \Delta(\underset{\sim}{x}) = 0, \tag{4.17}$$

or

$$\begin{aligned} \frac{d}{dt} V^*(\underset{\sim}{x}) &= (\nabla_{\underset{\sim}{y}} V)^T \; \underset{\approx}{J} \frac{d}{dt} \underset{\sim}{x} \\ &= - (\Delta(\underset{\sim}{x}))^2 (\nabla_{\underset{\sim}{y}} V)^T \; \underset{\sim}{F}^* \le (\Delta(\underset{\sim}{x}))^2 \; K(\nabla_{\underset{\sim}{y}} V) \le 0 \; . \end{aligned} \tag{4.18}$$

It thus follows that every solution of the system (4.14) will have its large time limit point in the set $\chi \cup \mathcal{D}$. In addition, since $V^*(\underset{\sim}{x})$ achieves its absolute minimum of zero on the set χ (i.e., $V^*(\underset{\sim}{x}) = V(\underset{\sim}{y}(\underset{\sim}{x}))$), each isolated point $\bar{\underset{\sim}{x}}$ of χ which does not belong to $\mathcal{D}$ is contained as an interior point of a neighborhood $N(\bar{\underset{\sim}{x}})$ that contains no points of $\mathcal{D}$, and for which every solution of (4.14) which starts in $N(\bar{\underset{\sim}{x}})$ will have $\bar{\underset{\sim}{x}}$ as its large time limit point.

Last, we come to the question of what measures are to be taken in the event that the given functions $\underset{\sim}{y}(\underset{\sim}{x})$ do not have

the property that $\| \underset{\sim}{x} \| \to \infty$ implies $\| \underset{\sim}{y}(x) \| \to \infty$. The crux
of the matter is that the function $V^*(\underset{\sim}{x}) = V(\underset{\sim}{y}(x))$ can now
decrease as $\| \underset{\sim}{x} \| \to \infty$ in one or more directions. For example,
$V(\underset{\sim}{y}) = (\underset{\sim}{y},\underset{\sim}{y})$, $\underset{\sim}{y} = \underset{\sim}{x}e^{-(\underset{\sim}{x},\underset{\sim}{x})}$, $V^2(\underset{\sim}{x}) = (\underset{\sim}{x},\underset{\sim}{x})e^{-2(\underset{\sim}{x},\underset{\sim}{x})}$. In this
circumstance, since $V^*(\underset{\sim}{x}(t))$ is nonincreasing along every orbit
of either the system (4.6) or (4.14), we are unable to preclude
the possibility of $\| \underset{\sim}{x}(t) \| \to \infty$ as t→∞ for certain initial
data. Let S be a smooth (N-1)-dimensional closed, oriented and
simply connected surface in E_N and let $\underset{\sim}{n}$ be the field of out-
ward oriented unit normal vectors to S in E_N. If

$$(\underset{\sim}{n}, \underset{\sim}{\nabla}_x V^*)\Big|_S < 0 \qquad (4.19)$$

holds at every point of S, then $V^*(\underset{\sim}{x})$ decreases as we move
away from S into the interior of the region bounded by S. It
thus follows, since $V^*(\underset{\sim}{x})$ decreases along all orbits of either
(4.6) or (4.14), that every solution to either (4.6) or (4.14)
with initial data in the interior of the region bounded by S will
have its large time limit point in the interior of the region
bounded by S. Thus satisfaction of the condition (4.19) will
assure the desired property even if $\| \underset{\sim}{x} \| \to \infty$ does not imply
$\| \underset{\sim}{y}(x) \| \to \infty$. This test is particularly useful when interest
lies only in zeros of $\underset{\sim}{y}(x)$ for $\underset{\sim}{x}$ in some given region of E_N,
or when $\underset{\sim}{y}(x)$ is only defined for $\underset{\sim}{x}$ in some given region of
E_N. An example of the latter situation arises in attempts to
find equilibrium concentrations of reacting mixtures; that is,
in solving the nonlinear laws of mass action and the linear laws
of mass balance for mass-closed chemical systems.

5. EXAMPLES AND CONVERGENCE RATES

We construct specific examples in this Section for which the
rate of convergence of the methods can be determined. Although we
confine our analysis to the system studied in Section 2, it is
evident that these same considerations can be applied to the convergence problem for either the system (4.6) or (4.14).

We take

$$V(\underset{\sim}{y}) = \frac{1}{2}(\underset{\sim}{y},\underset{\sim}{y}) = \frac{1}{2}\| \underset{\sim}{y} \|^2 , \tag{5.1}$$

so that $V^*(\underset{\sim}{x}) = V(\underset{\sim}{y}(\underset{\sim}{x}))$ is proportional to the square of the error in solving $\underset{\sim}{y}(\underset{\sim}{x}) = \underset{\sim}{0}$ and (2.5) gives

$$\underset{\sim}{W}(\underset{\sim}{y}) = \underset{\sim}{y} . \tag{5.2}$$

Accordingly, if we select $\underset{\sim}{F}(\underset{\sim}{y})$ in accordance with the asymptotic stability theorem given in Section 2, and take $\underset{\sim}{U}(\underset{\sim}{y}) = 0$, then (2.2), (2.5) and (5.2) give

$$\underset{\sim}{F}(\underset{\sim}{y}) = \nabla_{\underset{\sim}{y}}\phi(\underset{\sim}{y}) \tag{5.3}$$

and we have

$$\frac{d}{dt}\underset{\sim}{y} = - \nabla_{\underset{\sim}{y}}\phi(\underset{\sim}{y}) = - \nabla_{\underset{\sim}{y}} \int_o^1 \{P(\lambda\underset{\sim}{y}) - K(\lambda\underset{\sim}{y})\} \frac{d\lambda}{\lambda} \tag{5.4}$$

$$\frac{d}{dt} V = K(\underset{\sim}{y}) - P(\underset{\sim}{y}) \leq K(\underset{\sim}{y}) \leq 0 \tag{5.5}$$

since (2.9) - (2.13) hold. Since $K(\underset{\sim}{y})$ is at our disposal, subject only to the conditions $K(\underset{\sim}{y}) \leq 0$, $K(\underset{\sim}{y}) = 0 \Leftrightarrow \underset{\sim}{y} = \underset{\sim}{0}$, we are free to select $K(\underset{\sim}{y})$ by

$$K(\underset{\sim}{y}) = - Q(\tfrac{1}{2}\| \underset{\sim}{y} \|^2) = - Q(V) , \tag{5.6}$$

where $Q(V)$ in a C^2 function of its argument such that

$$Q(V) \geq 0 \quad \text{for} \quad V \geq 0 , \quad Q(V) = 0 \Longleftrightarrow V = 0 . \tag{5.7}$$

In this event, (5.5) becomes

$$\frac{d}{dt} V = - Q(V) - P(\underset{\sim}{y}) \leq - Q(V) \tag{5.8}$$

since $P(\underset{\sim}{y}) \geq 0$. It thus follows that every solution of (5.4) will have the property

$$\| \underset{\sim}{y}(t) \|^2 \leq 2v(t) , \tag{5.9}$$

where $v(t)$ satisfies the differential equation

$$\frac{dv}{dt} = - Q(v) \tag{5.10}$$

subject to the initial condition

$$v(0) = \frac{1}{2}\| \underset{\sim}{y}(0) \|^2 . \tag{5.11}$$

We note in passing that equality obtains in (5.9) if and only if we choose the function $P(\underset{\sim}{y})$ such that $P(\underset{\sim}{y}) \equiv 0$.

It is now a straightforward problem to choose the function $Q(V)$ in such a way to obtain various rates of convergence. An obvious first choice is

$$Q(V) = k\,V , \quad k > 0 , \tag{5.12}$$

which obviously satisfies the conditions (5.7). In this case, (5.10) becomes $dv/dt = kv$, and hence (5.9) and (5.11) yield

$$\| \underset{\sim}{y}(t) \|^2 \leq \| \underset{\sim}{y}(0) \|^2 \exp(-kt) ; \tag{5.13}$$

that is, we have exponential convergence. In particular, if we take $P(\underset{\sim}{y}) \equiv 0$, then $\underset{\sim}{F}(\underset{\sim}{y}) = -k\underset{\sim}{y}$ and we obtain strict equality in (5.13). Application of these results to the problem of solving $\underset{\sim}{y}(\underset{\sim}{x}) = \underset{\sim}{0}$ by means of the system (4.14) and (4.15) gives

$$\frac{d}{dt}\underset{\sim}{x}(t) = -\Delta(\underset{\sim}{x})\, \underset{\approx}{C}(\underset{\sim}{x})^T\, \underset{\sim}{y}(\underset{\sim}{x}) , \tag{5.14}$$

$$\frac{d}{dt}\underset{\sim}{y}(\underset{\sim}{x}(t)) = -(\Delta(\underset{\sim}{x}))^2\, \underset{\sim}{y}(\underset{\sim}{x}) . \tag{5.15}$$

The system (5.15) is seen to be quite similar to the Davidenko-Branin [3] method, $d\underset{\sim}{y}(\underset{\sim}{x})/dt = \text{sign}(\Delta(\underset{\sim}{x}))\underset{\sim}{y}$, with the exception that (5.15) has a C^1 right-handed side. The function $\text{sign}\,(\Delta(\underset{\sim}{x}))$ which occurs in the Davidenko system has a jump discontinuity on the set $\mathcal{D}$, and all of the attendant troubles which such discontinuities cause, not to mention the problems associated with $\Delta(\underset{\sim}{x}) = 0$.

A more striking example is that provided by the choice

$$Q(V) = k\,V(1+(\ell nV)^2) , \; k > 0 \tag{5.16}$$

which clearly satisfies the conditions (5.7). In this case, (5.10) gives

$$\frac{dv}{dt} = -\ k\ v(1+\ell n v)^2) \ , \tag{5.17}$$

and hence (5.11) and (5.17) yield

$$v(t) = \exp(-\tan(kt+c)) \tag{5.18}$$

$$-\ \pi/2 \leq c = -\ \tan^{-1}(\ell n(\tfrac{1}{2}\| \underset{\sim}{y}(0)\|^2) \leq \pi/2 \ . \tag{5.19}$$

Thus, (5.9) shows that

$$\| \underset{\sim}{y}(t)\|^2 \leq 2\ \exp(-\tan(kt+c)) \ ; \tag{5.20}$$

that is, *we obtain exponential convergence in finite time* $T = (\frac{\pi}{2} - c)k^{-1}$ where c is given by (5.19).

If we wish to apply this result to the problem of solving $\underset{\sim}{y}(\underset{\sim}{x}) = \underset{\sim}{0}$ by means of the systems (4.14) and (4.15), then we must compute $\underset{\sim}{F}(\underset{\sim}{y})$, and hence $\phi(\underset{\sim}{y})$. Now for $P(\underset{\sim}{y}) = 0$,

$$\phi(y) = -\int_0^1 K(\lambda \underset{\sim}{y})\ \frac{d\lambda}{\lambda} = +\int_0^1 Q(\lambda^2 V)\ \frac{d\lambda}{\lambda}$$

$$= kV \int_0^1 \{1+(\ell n\lambda^2 V)^2\}\lambda d\lambda$$

$$= \frac{k}{2}\ V\{3-2\ell n V + (\ell n V)^2\}$$

and hence (5.3) gives

$$\underset{\sim}{F}(\underset{\sim}{y}) = \frac{k}{2}\ \{1+(\ell n V)^2\}\underset{\sim}{y} \ . \tag{5.21}$$

Thus (4.14) and (4.15) yield

$$\frac{d}{dt}\ \underset{\sim}{x}(t) = -\ \frac{k\Delta(\underset{\sim}{x})}{2}\ \{1+(\ell n V)^2\}\underset{\approx}{C}(\underset{\sim}{x})^T\ \underset{\sim}{y}(\underset{\sim}{x}) \ , \tag{5.22}$$

$$\frac{d}{dt}\ \underset{\sim}{y}(\underset{\sim}{x}) = -\ \frac{k}{2}\ (\Delta(\underset{\sim}{x}))^2\{1+(\ell n V)^2\}\underset{\sim}{y}(\underset{\sim}{x}) \tag{5.23}$$

and

$$\frac{dV}{dt} = (\underset{\sim}{y},\ \frac{d\underset{\sim}{y}}{dt}) = -\ k(\Delta(\underset{\sim}{x}))^2\{1+(\ell n V)^2\}V \ . \tag{5.24}$$

Thus, if the point set $N \equiv \{\underset{\sim}{x}\varepsilon E_N |\ V(\underset{\sim}{y}(\underset{\sim}{x})) = \frac{1}{2}\| \underset{\sim}{y}\|^2 < R^2\}$ contains no points such that $\Delta(\underset{\sim}{x}) = 0$, then there exists a number $\delta > 0$ such that

$$0 < \sqrt{\delta} = \min_{\underset{\sim}{x} \in N} (\Delta(\underset{\sim}{x})) \tag{5.25}$$

and (5.24) yields

$$\frac{dV}{dt} \leq - \delta kV\{1+(\ell nV)^2\} \leq 0 . \tag{5.26}$$

Hence, every solution of (5.22) which starts in N will converge such that

$$\| \underset{\sim}{y}(\underset{\sim}{x}) \|^2 \leq 2 \exp(-\tan(\delta kt+c)) \tag{5.27}$$

for

$$c = - \tan^{-1}(\ell n(\tfrac{1}{2}\| \underset{\sim}{y}(0) \|^2) . \tag{5.28}$$

REFERENCES

1. Eells, James, Jr. and J.H. Sampson: *Amer. J. Math. 86,* 109 (1964).
2. Hartman, P.: *Canadian J. Math. 19,* 673 (1967).
3. Hahn, W.: *Stability of Motion* (Springer, Berlin, 1967)
4. Edelen, D.G.B.: *Arch. Rat'l Mech. Anal. 51,* 218 (1973); *Int. J. Engng. Sci. 12,* 121 (1974).
5. Davidenko, D.F.: *Ukr. Mat. Z. 5,* 196 (1953); Branin, F.H.: *Memoirs IEEE Conference on Systems, Nerworks and Computers* (Oaxtepec, Mexico, 1971).

Differential Procedures for Systems of Implicit Relations and Implicitly Coupled Nonlinear Boundary-Value Problems

Dominic G. B. Edelen

1. INTRODUCTION AND PRELIMINARY CONSIDERATIONS

Theory, in the form of the implicit function theorem, is quite specific about solvability of implicit systems of relations. Numerical procedures for the realization of the solutions, when the conditions of the implicit function theorem are met, is quite another matter. The purpose of this note is to give families of differential procedures for solving systems of implicit relations.

Let $\underset{\sim}{x}$ denote a vector (column matrix) in n-dimensional number space E_n and let $\underset{\sim}{\alpha}$ denote a vector in m-dimensional number space E_m. Let $\underset{\sim}{f}(\underset{\sim}{x},\underset{\sim}{\alpha})$ be a given vector valued function that is defined for all $\underset{\sim}{x}$ in a given n-dimensional region R_n of E_n and all $\underset{\sim}{\alpha}$ in E_m and which takes its values in E_m. Thus, $\underset{\sim}{f}(\underset{\sim}{x},\underset{\sim}{\alpha})$ is a mapping of $R_n \times E_m$ into E_m. We further assume that $\underset{\sim}{f}$ is continuous over its domain of definition and that its matrices of partial derivatives

$$\underset{\sim}{A}(\underset{\sim}{x},\underset{\sim}{\alpha}) = \underset{\sim}{\nabla}_x \underset{\sim}{f} \ , \quad \underset{\sim}{B}(\underset{\sim}{x},\underset{\sim}{\alpha}) = \underset{\sim}{\nabla}_\alpha \underset{\sim}{f} \tag{1.1}$$

are also continuous on the domain of definition of $\underset{\sim}{f}$. To be more specific, let i, j, k be indices which can take on values from 1 through n and let a, b, c be indices that can take on values from 1 through m. We then have

$$\underset{\sim}{A} = ((\partial f_a/\partial x_i)) \ , \quad \underset{\sim}{B} = ((\partial f_a/\partial \alpha_b)) \ ,$$

so that $\underset{\sim}{A}$ is an m-by-n matrix and $\underset{\sim}{B}$ is an m-by-m matrix.

The problem we wish to solve is that of constructing differential procedures for obtaining $\underset{\sim}{\alpha} = \underset{\sim}{\phi}(\underset{\sim}{x})$ as a vector valued function of $\underset{\sim}{x}$ such that

$$\underset{\sim}{f}(\underset{\sim}{x},\underset{\sim}{\phi}(\underset{\sim}{x})) = \underset{\sim}{0} \tag{1.2}$$

is satisfied at all points $\underset{\sim}{x}$ of E_n for which such solutions exist. The implicit function theorem tells us that if there are elements $\bar{\underset{\sim}{x}}$ and $\bar{\underset{\sim}{\alpha}}$ such that

$$\underset{\sim}{f}(\bar{\underset{\sim}{x}},\bar{\underset{\sim}{\alpha}}) = \underset{\sim}{0} \tag{1.3}$$

and $\det(\underset{\sim}{B}(\bar{\underset{\sim}{x}},\bar{\underset{\sim}{\alpha}})) \neq 0$, then there exists an n-dimensional neighborhood $N(\bar{\underset{\sim}{x}})$ of $\bar{\underset{\sim}{x}}$ in R_n such that $\underset{\sim}{\alpha} = \underset{\sim}{\phi}(\underset{\sim}{x})$ exists on $N(\bar{\underset{\sim}{x}})$ and $\underset{\sim}{f}(\underset{\sim}{x},\underset{\sim}{\phi}(\underset{\sim}{x})) = \underset{\sim}{0}$ is indentically satisfied for all $\underset{\sim}{x}$ in $N(\bar{\underset{\sim}{x}})$. This neighborhood $N(\bar{x})$ has the additional property that $\det(\underset{\sim}{B}(\underset{\sim}{x},\underset{\sim}{\phi}(\underset{\sim}{x})) \neq 0$. Further, for all continuously differentiable functions $\underset{\sim}{x}(t)$ with range in $N(\bar{\underset{\sim}{x}})$, we have

$$\underset{\sim}{A}\frac{d\underset{\sim}{x}}{dt} + \underset{\sim}{B}\frac{d\overset{*}{\underset{\sim}{\phi}}}{dt} = \underset{\sim}{0} \tag{1.4}$$

(matrix multiplication); that is, $d\overset{*}{\underset{\sim}{f}}/dt = 0$. Here, we have set

$$\overset{*}{\underset{\sim}{\phi}}(t) = \underset{\sim}{\phi}(\underset{\sim}{x}(t)) \ , \quad \overset{*}{\underset{\sim}{f}}(t) = \underset{\sim}{f}(\underset{\sim}{x}(t), \overset{*}{\underset{\sim}{\phi}}(t)) \ . \tag{1.5}$$

Clearly, there are three numerical problems. The first is that of finding two numerical vectors $\bar{\underset{\sim}{x}}$ and $\bar{\underset{\sim}{\alpha}}$ such that $\underset{\sim}{f}(\bar{\underset{\sim}{x}},\bar{\underset{\sim}{\alpha}}) = \underset{\sim}{0}$ and $\det(\underset{\sim}{B}(\bar{\underset{\sim}{x}},\bar{\underset{\sim}{\alpha}})) \neq 0$. The second is that of finding values of $\underset{\sim}{\phi}(\underset{\sim}{x})$ for given values of $\underset{\sim}{x}$ in $N(\bar{\underset{\sim}{x}})$ such that $\underset{\sim}{\alpha} = \underset{\sim}{\phi}(\underset{\sim}{x})$ satisfies $\underset{\sim}{f}(\underset{\sim}{x},\underset{\sim}{\alpha}) = \underset{\sim}{0}$. The third problem is that of finding other pairs of numerical vectors $\bar{\underset{\sim}{y}}$ and $\bar{\underset{\sim}{\beta}}$ such that $\underset{\sim}{f}(\bar{\underset{\sim}{y}},\bar{\underset{\sim}{\beta}}) = \underset{\sim}{0}$, $\det(\underset{\sim}{B}(\bar{\underset{\sim}{y}},\bar{\underset{\sim}{\beta}})) \neq 0$, and the pair $(\bar{\underset{\sim}{y}},\bar{\underset{\sim}{\beta}})$ does not belong to the solution set obtained by starting with the pair $(\bar{\underset{\sim}{x}},\bar{\underset{\sim}{\alpha}})$. The possibility of the existence of such additional pairs

$(\bar{y}, \bar{\beta})$ can not be excluded since the implicit function theorem only gives sufficient conditions for the existence of local single-valued solutions. For instance, consider the problem of solving $x^2 + \alpha^2 - r^2 = 0$. Clearly there are two solutions $\alpha = \pm \sqrt{r^2 - x^2}$ for each x in the open interval $(-r, r)$ and only one of these can be obtained by starting with the pair of points $x = 0$, $\alpha = r$.

The second problem is treated first in Section 2 since it is the easier of the three. The first problem is then solved in Section 3 and Section 4 takes up the third problem.

One of the most vexing classes of problems in which implicit relations occur consists of nonlinear boundary value problems in which there is coupling that is defined through a collection of implicit relations. Section 5 gives a method of solution for such problems by converting them to initial value problems in one additional variable t such that the large time limit of the solution of the initial value problem solves the given boundary value problem. Such a procedure is particularly useful in that the initial data can be chosen as any convenient vector of functions that satisfies the given boundary conditions.

2.   GENERATION OF A SOLUTION ON A GRID IN E_n

We assume in this Section that we know a pair of numerical vectors $\bar{x}$ and $\bar{\alpha}$ such that $\underset{\sim}{f}(\bar{x}, \bar{\alpha}) = \underset{\sim}{0}$ and $\det(B(\bar{x}, \bar{\alpha})) \neq 0$. Let $\underset{\sim}{e}_i$, $i = 1, \ldots, n$, be an orthonormal system of basis vectors for E_n. The set of all points

$$\underset{\sim}{x}_1(t) = \bar{x} + t \underset{\sim}{e}_1 \tag{2.1}$$

defines a line in E_n that passes through the point $\bar{x}$ and is parallel to the vector $\underset{\sim}{e}_1$. It is therefore meaningful to ask for a vector function $\underset{\sim}{\alpha}_1(t)$ such that

$$\underset{\sim}{f}(\underset{\sim}{x}_1(t), \underset{\sim}{\alpha}_1(t)) = \underset{\sim}{0} \tag{2.2}$$

$$\underset{\sim}{\alpha}_1(0) = \bar{\alpha} . \tag{2.3}$$

When (2.1) is used to calculate the time derivative of (2.2), we obtain the system of differential equations

$$\underset{\sim}{B}(\bar{\underset{\sim}{x}} + t\underset{\sim}{e}_1, \underset{\sim}{\alpha}_1(t)) \frac{d\underset{\sim}{\alpha}_1}{dt} = - \underset{\sim}{A}(\bar{\underset{\sim}{x}} + t\underset{\sim}{e}_1, \underset{\sim}{\alpha}_1) \underset{\sim}{e}_1 \qquad (2.4)$$

and the initial data (2.3) for the determination of the vector $\underset{\sim}{\alpha}_1(t)$. Since $\det(\underset{\sim}{B}(\bar{\underset{\sim}{x}}, \bar{\underset{\sim}{\alpha}})) \neq 0$ and $\underset{\sim}{B}$ is a continuous matrix-valued function of its arguments, it follows that there is a maximal open interval n_1 that contains the point $t = 0$ and is such that a solution exists to (2.4) subject to the initial date (2.2) for all t in n_1. Clearly, $\det(\underset{\sim}{B}(\bar{\underset{\sim}{x}} + t\underset{\sim}{e}_1, \underset{\sim}{\alpha}_1(t))) \neq 0$ for t in n_1, and hence the system (2.4) can be solved for the vector $d\underset{\sim}{\alpha}_1(t)/dt$ and then integrated numerically. The reason that we can not guarantee existence of solutions to the system (2.4) for all t is that $\det(\underset{\sim}{B}(\underset{\sim}{x} + t\underset{\sim}{e}_1, \underset{\sim}{\alpha}(t)))$ can go to zero as t approaches the endpoints of the closure of n_1. The solution $\underset{\sim}{\alpha}_1(t)$ allows us to generate a mesh of pairs of values by

$$\underset{\sim}{x}_k = \bar{\underset{\sim}{x}} + k \, h \, \underset{\sim}{e}_1 \ , \qquad (2.5)$$

$$\underset{\sim}{\alpha}_k = \underset{\sim}{\alpha}_1(k \, h) \ , \qquad (2.6)$$

where h is a given constant (mesh constant) and k ranges over the integers such that kh belongs to n_1.

We now start with each of the points $\bar{\underset{\sim}{x}} + k \, h \, \underset{\sim}{e}_1$ and generate the line

$$\underset{\sim}{x}_{2k}(t) = \bar{\underset{\sim}{x}} + kh\underset{\sim}{e}_1 + t\underset{\sim}{e}_2 \ . \qquad (2.7)$$

Since $\det(\underset{\sim}{B}(\bar{\underset{\sim}{x}} + kh\underset{\sim}{e}_1, \underset{\sim}{\alpha}_1(kh))) \neq 0$, we can determine vector function $\underset{\sim}{\alpha}_{2k}(t)$ for each value of k by solving the system of differential equations

$$\underset{\sim}{B}(\bar{\underset{\sim}{x}} + kh\underset{\sim}{e}_1 + t\underset{\sim}{e}_2, \underset{\sim}{\alpha}_{2k}(t)) \frac{d\underset{\sim}{\alpha}_{2k}}{dt} = - \underset{\sim}{A}(\bar{\underset{\sim}{x}} + kh\underset{\sim}{e}_1 + t\underset{\sim}{e}_2, \underset{\sim}{\alpha}_{2k}(t))\underset{\sim}{e}_2 \quad (2.8)$$

subject to the initial data

$$\underset{\sim}{\alpha}_{2k}(0) = \underset{\sim}{\alpha}_k = \underset{\sim}{\alpha}_1(kh) \ , \qquad (2.9)$$

for all t in a maximal open interval n_{2k} that contains $t = 0$.
As before, we construct the two-dimensional mesh of pairs

$$x_{kj} = \bar{x} + kh\underset{\sim}{e}_1 + jh\underset{\sim}{e}_2 , \qquad (2.10)$$

$$\alpha_{kj} = \alpha_{2k}(jh) , \qquad (2.11)$$

by allowing t to take on the values $j\,h$ for j ranging over
the integers such that $j\,h$ belongs to n_{2k}

This procedure can be continued by successive use of the
basis vectors $\underset{\sim}{e}_3, \underset{\sim}{e}_4, \ldots$ until a mesh of pairs of numerical
vectors $\underset{\sim}{x}$ and $\underset{\sim}{\alpha}$ is built up such that the mesh of numerical
vectors $\underset{\sim}{x}$ constitutes a set of grid points on E_n with grid
spacing h that spans an n-dimensional region R_n of E_n.
Clearly, the mesh so generated constitutes a numerical solution of
$\underset{\sim}{f}(x,\alpha) = \underset{\sim}{0}$ on the region R_n. It is also clear that the region
R_n is maximal since each of the open intervals n_1, n_{2k}, n_{3kj},
etc. is maximal.

3. GENERATION OF THE STARTING VALUES FOR THE GRID

The problem we now have to solve is that of finding a pair
of numerical vectors $\underset{\sim}{\bar{x}}$ and $\underset{\sim}{\bar{\alpha}}$ such that $\underset{\sim}{f}(\bar{x},\bar{\alpha}) = 0$.

We start by picking a specific numerical vector $\underset{\sim}{\bar{x}}$. It is
then useful to introduce the simplifying notation

$$\underset{\sim}{\bar{f}}(\alpha) = \underset{\sim}{f}(\bar{x},\alpha) , \qquad (3.1)$$

$$\underset{\sim}{\bar{B}}(\alpha) = \underset{\sim}{B}(\bar{x},\alpha) . \qquad (3.2)$$

In these terms, we need to find a vector $\underset{\sim}{\alpha}$ such that

$$\underset{\sim}{\bar{f}}(\alpha) = \underset{\sim}{0} \qquad (3.3)$$

If there is *a priori* information to the effect that $\det(\bar{B}(\underset{\sim}{\alpha})) \neq 0$
for all $\underset{\sim}{\alpha}$ in E_m, then an intrinsic economy can be achieved in
the solution of (3.3). We therefore divide our considerations
into two separate cases

Case I. $\mathrm{Det}(\bar{B}(\underset{\sim}{\alpha})) \neq 0$ *for all* $\underset{\sim}{\alpha}$.

In this case, we can use the Davidenko method [1]. This method consists of picking an arbitrary initial vector $\underset{\sim}{\alpha}_o$ for $\underset{\sim}{\alpha}$ and solving the system of autonomous differential equations

$$\frac{d\bar{f}(\underset{\sim}{\alpha})}{dt} = \bar{\underset{\sim}{B}}(\underset{\sim}{\alpha}) \frac{d\underset{\sim}{\alpha}}{dt} = - K \bar{\underset{\sim}{f}}(\underset{\sim}{\alpha}) , \tag{3.4}$$

where K is a positive constant. Since $\det(\underset{\sim}{B}) \neq 0$ for all $\underset{\sim}{\alpha}$, this system is equivalent to

$$\frac{d\underset{\sim}{\alpha}}{dt} = - K(\bar{\underset{\sim}{B}})^{-1} \bar{\underset{\sim}{f}}(\underset{\sim}{\alpha}) \tag{3.5}$$

where $(\bar{\underset{\sim}{B}})^{-1}$ is the inverse of the matrix $\bar{\underset{\sim}{B}}$. It then follows, on setting

$$V(\underset{\sim}{\alpha}) = \frac{1}{2} \bar{\underset{\sim}{f}}(\underset{\sim}{\alpha})^T \bar{\underset{\sim}{f}}(\underset{\sim}{\alpha}) \tag{3.6}$$

($\bar{\underset{\sim}{f}}^T$ = transpose of $\bar{\underset{\sim}{f}}$) that

$$\frac{dV}{dt} = \bar{\underset{\sim}{f}}^T \underset{\sim}{B} \frac{d\underset{\sim}{\alpha}}{dt} = - K\underset{\sim}{f}^T\underset{\sim}{f} = - 2KV , \tag{3.7}$$

and hence

$$\overset{*}{V}(t) = \frac{1}{2} \bar{f}(\underset{\sim}{\alpha}(t))^T \bar{f}(\underset{\sim}{\alpha}(t)) = V_o \exp(-2Kt) . \tag{3.8}$$

Accordingly, $\lim_{t\to\infty}\{\underset{\sim}{\alpha}(t)\} = \bar{\alpha}$ exists and satisfies $\bar{\underset{\sim}{f}}(\bar{\underset{\sim}{\alpha}}) = \underset{\sim}{0}$. This solves the problem for we then have the required pair of vectors $\bar{\underset{\sim}{x}}$ and $\bar{\underset{\sim}{\alpha}}$ such that $\underset{\sim}{f}(\bar{\underset{\sim}{x}},\bar{\underset{\sim}{\alpha}}) = \underset{\sim}{0}$.

It is noted in passing that $\bar{\underset{\sim}{\alpha}}$ is the only vector such that $\underset{\sim}{f}(\bar{\underset{\sim}{x}},\bar{\underset{\sim}{\alpha}}) = \underset{\sim}{0}$ for the given numerical vector $\bar{\underset{\sim}{x}}$. This follows from the fact that $\det(\bar{\underset{\sim}{B}}(\underset{\sim}{\alpha})) \neq 0$ for all $\underset{\sim}{\alpha}$ is sufficient in order to insure that the mapping $\underset{\sim}{\beta} = \bar{\underset{\sim}{f}}(\underset{\sim}{\alpha})$ is a one-to-one mapping of E_m to E_m.

Case II. No A Priori Information about $\det(\bar{\underset{\sim}{B}})$.

Clearly, the Davidenko method can not be used in this case since the coefficient matrix $\bar{\underset{\sim}{B}}(\underset{\sim}{\alpha})$ of the differential system (3.4) may become singular for some values of the vector $\underset{\sim}{\alpha}$. Instead of the system (3.4), we now use the autonomous differential system

$$\frac{d\underset{\sim}{\alpha}}{dt} = - K\, \underset{\sim}{\bar{B}}(\underset{\sim}{\alpha})^T\, \underset{\sim}{\bar{f}}(\underset{\sim}{\alpha}) \tag{3.9}$$

where K is a positive constant. If we assume that each of the entries of the matrix $\underset{\sim}{\bar{B}}(\underset{\sim}{\alpha})$ is a C^1 function of the vector $\underset{\sim}{\alpha}$, the right hand sides of (3.9) are C^1 functions of $\underset{\sim}{\alpha}$ so that there is no question about existence of solutions. Further, with $V(\underset{\sim}{\alpha})$ defined by (3.6), the system (3.9) implies that

$$\frac{dV}{dt} = \underset{\sim}{\bar{f}}^T \underset{\sim}{\bar{B}} \frac{d\underset{\sim}{\alpha}}{dt} = - K \underset{\sim}{\bar{f}}^T \underset{\sim}{\bar{B}} \underset{\sim}{\bar{B}}^T \underset{\sim}{\bar{f}} = \left(\frac{-1}{K}\right) \frac{d\underset{\sim}{\alpha}}{dt}^T \frac{d\underset{\sim}{\alpha}}{dt} \le 0 \tag{3.10}$$

with $dV/dt = 0$ if and only if $\underset{\sim}{\bar{B}}^T \underset{\sim}{f} = \frac{d\underset{\sim}{\alpha}}{dt} = \underset{\sim}{0}$. Accordingly, $V(\underset{\sim}{\alpha}(t))$ decreases along every solution of the system (3.9) whenever $d\underset{\sim}{\alpha}/dt \neq 0$. Since $V \geq 0$ with $V(\underset{\sim}{\alpha}) = 0$ if and only if $\underset{\sim}{\bar{f}}(\underset{\sim}{\alpha}) = \underset{\sim}{0}$, every solution of the system (3.9), with an initial value of $\underset{\sim}{\alpha}$ such that $d\underset{\sim}{\alpha}(0)/dt \neq \underset{\sim}{0}$, will converge for large values of t to values of $\underset{\sim}{\alpha}$ for which either $V(\underset{\sim}{\alpha}) = 0$ or $V(\underset{\sim}{\alpha}) =$ relative minimum. For those large time limit points such that $V(\underset{\sim}{\alpha}) = 0$, we have $\underset{\sim}{\bar{f}}(\underset{\sim}{\alpha}) = \underset{\sim}{0}$ and the problem is solved. For those large time limit points such that $V(\underset{\sim}{\alpha}) \neq 0$, we have $\underset{\sim}{\bar{B}}^T(\underset{\sim}{\alpha})\, \underset{\sim}{\bar{f}}(\underset{\sim}{\alpha}) = \underset{\sim}{0}$ but $\underset{\sim}{\bar{f}}(\underset{\sim}{\alpha}) \neq \underset{\sim}{0}$, and hence such points satisfy $\det(\underset{\sim}{\bar{B}}(\underset{\sim}{\alpha})) = 0$. We have thus established that the large time limit points of the system (3.9) exist for all initial data and contain the set of all points $\underset{\sim}{\alpha}$ for which $\underset{\sim}{\bar{f}}(\underset{\sim}{\alpha}) = \underset{\sim}{0}$. We also obtain spurious convergence points where $\underset{\sim}{\bar{B}}^T(\underset{\sim}{\alpha})\, \underset{\sim}{\bar{f}}(\underset{\sim}{\alpha}) = \underset{\sim}{0}$, but these are easily eliminated by testing to see whether the large time limit points satisfy $\underset{\sim}{f}(\underset{\sim}{\alpha}) = \underset{\sim}{0}$. This testing procedure is a simple one since the value of the vector $\underset{\sim}{\alpha}$ to be tested is known.

We note in passing that $V(\underset{\sim}{\alpha})$ is proportional to the square of the error in solving $\underset{\sim}{\bar{f}}(\underset{\sim}{\alpha}) = \underset{\sim}{0}$, and that the system (3.9) can be written in the equivalent form

$$\frac{d\underset{\sim}{\alpha}}{dt} = - K\, \underset{\sim}{\nabla}_{\alpha} V(\underset{\sim}{\alpha}) \ . \tag{3.11}$$

Thus, the differential procedure (3.9) is the differential analog of steepest descent methods; that is, $d\underset{\sim}{\alpha}/dt$ is a vector that points in the direction of maximal decrease of the error.

4. MULTIPLE BRANCH SOLUTIONS

If we know that $\det(\underset{\sim}{B}(\bar{\underset{\sim}{x}},\underset{\sim}{\alpha})) \neq 0$ for all $\underset{\sim}{\alpha}$, then we have seen that there is only one vector $\bar{\underset{\sim}{\alpha}}$ such that $\underset{\sim}{f}(\bar{\underset{\sim}{x}},\bar{\underset{\sim}{\alpha}}) = \underset{\sim}{0}$. Thus, there is one and only one solution of $\underset{\sim}{f}(\underset{\sim}{x},\underset{\sim}{\alpha}) = 0$ over the point $\underset{\sim}{x} = \bar{\underset{\sim}{x}}$. On the other hand, if $\det(\underset{\sim}{B}(\bar{\underset{\sim}{x}},\underset{\sim}{\alpha}))$ can vanish for some vector $\underset{\sim}{\alpha}$, there can be more than one solution of $\underset{\sim}{f}(\bar{\underset{\sim}{x}},\underset{\sim}{\alpha})=\underset{\sim}{0}$. This follows from the results established under Case II of the previous section. If there are several solutions $\bar{\underset{\sim}{\alpha}}_1, \bar{\underset{\sim}{\alpha}}_2,\ldots,\bar{\underset{\sim}{\alpha}}_p$ of $\underset{\sim}{f}(\bar{x},\underset{\sim}{\alpha}) = \underset{\sim}{0}$ that are obtained from the differential procedure (3.9) by allowing the initial $\underset{\sim}{\alpha}$ vector to change, we can use the method described in Section 2 to generate a solution grid for each starting pair $(\bar{\underset{\sim}{x}},\bar{\underset{\sim}{\alpha}}_1)$, $(\bar{\underset{\sim}{x}},\bar{\underset{\sim}{\alpha}}_2),\ldots,$ $(\bar{\underset{\sim}{x}},\bar{\underset{\sim}{\alpha}}_p)$. Clearly, in this instance, $\underset{\sim}{f}(\underset{\sim}{x},\underset{\sim}{\alpha}) = \underset{\sim}{0}$ has p-fold valued solutions. It is also clear that the size of the grid over which each of these solution meshes is defined can be different for each different starting point, $(\bar{x},\bar{\alpha}_a)$. We do not consider in this note the very complex problems of whether the various value surfaces of the solution fit together where $\det(\underset{\sim}{B}) = 0$, the manner of this fitting together process, or the order in which the surfaces are fitted together.

5. NONLINEAR BOUNDARY VALUE PROBLEMS WITH IMPLICIT COUPLING

The results established above provide a means whereby differential procedures can be constructed for the solution of nonlinear boundary value problems with implicit coupling.

We again suppose that we are given the vector valued function $f(x;\underset{\sim}{\alpha})$ and that $\underset{\sim}{\alpha} = \underset{\sim}{\phi}(\underset{\sim}{x})$ is defined implicitly by

$$\underset{\sim}{f}(\underset{\sim}{x},\underset{\sim}{\alpha}) = \underset{\sim}{0} \,. \tag{5.1}$$

We further assume that the matrix $\underset{\sim}{B}(\underset{\sim}{x},\underset{\sim}{\alpha})$ defined by (1.1) is nonsingular so that

$$d\underset{\sim}{\alpha} = - \underset{\sim}{B}^{-1} \underset{\sim}{A} \, d\underset{\sim}{x} \tag{5.2}$$

holds for all vectors $\underset{\sim}{x}$. Now, consider the collection of vector valued functions $\underset{\sim}{x} = \underset{\sim}{x}(z,t)$ defined for all z in the closed interval $[a, b]$ and all t in $[0,\infty)$ and are such that

$$\underset{\sim}{x}(a,t) = \underset{\sim}{\gamma}_1 , \quad \underset{\sim}{x}(b,t) = \underset{\sim}{\gamma}_2 , \tag{5.3}$$

$\underset{\sim}{x}(z,t)$ has continuous second derivatives with respect to z , continuous first derivatives with respect to t and continuous mixed derivatives with respect to x and t . We wish to determine one such vector valued function which is such that it renders the functional

$$L(\underset{\sim}{x};\underset{\sim}{\alpha}) = \int_a^b \mathcal{L}(z,\underset{\sim}{x},\partial_z\underset{\sim}{x},\underset{\sim}{\alpha})dz \tag{5.4}$$

stationary in value. We assumed that $\mathcal{L}$ is an analytic function of its $2m + n + 1$ arguments such that

$$L(\underset{\sim}{x};\underset{\sim}{\alpha}) \geq J \tag{5.5}$$

for all $\underset{\sim}{x}(z,t)$ and all $\underset{\sim}{\alpha}(z,t)$ that are related by (5.1). Here $\partial_z\underset{\sim}{x}$ denotes the derivatives with respect to z . We shall also use $\partial_t\underset{\sim}{x}$ to denote the derivative with respect to t . The standard methods from the calculus of variations [2] yield

$$\delta L(\underset{\sim}{x};\underset{\sim}{\alpha}) = \int_a^b \left[\left\{\frac{\partial \mathcal{L}}{\partial \underset{\sim}{x}} - \partial_z\left(\frac{\partial \mathcal{L}}{\partial(\partial_z\underset{\sim}{x})}\right)\right\}\cdot\delta\underset{\sim}{x} + \frac{\partial \mathcal{L}}{\partial \underset{\sim}{\alpha}}\cdot\delta\underset{\sim}{\alpha}\right] dz. \tag{5.6}$$

If we define the row matrix of Euler-Lagrange derivatives by

$$\{E|\mathcal{L}\}_{\underset{\sim}{x}} = \frac{\partial \mathcal{L}}{\partial \underset{\sim}{x}} - \partial_z\left(\frac{\partial \mathcal{L}}{\partial(\partial_z\underset{\sim}{x})}\right) \tag{5.7}$$

and note that (5.2) implies $\delta\underset{\sim}{\alpha} = \underset{\sim}{B}^{-1} \underset{\sim}{A} \, \delta\underset{\sim}{x}$, (5.6) becomes

$$\delta L(\underset{\sim}{x};\underset{\sim}{\alpha}) = \int_a^b \left[\{E|\mathcal{L}\}_{\underset{\sim}{x}} - \frac{\partial \mathcal{L}}{\partial \underset{\sim}{\alpha}} \underset{\sim}{B}^{-1} \underset{\sim}{A}\right] \delta\underset{\sim}{x} \, dz , \tag{5.8}$$

so that we seek solutions to the implicit boundary value problem

$$\{E|L\}_{\underset{\sim}{x}} - \frac{\partial L}{\partial \underset{\sim}{\alpha}} \underset{\sim}{B}^{-1} \underset{\sim}{A} = \underset{\sim}{0}$$
$$f(\underset{\sim}{x},\underset{\sim}{\alpha}) = \underset{\sim}{0} \tag{5.9}$$

$$\underset{\sim}{x}(a,t) = \underset{\sim}{\gamma}_1 , \quad x(b,t) = \gamma_2 . \tag{5.10}$$

Now, as is well known, boundary value problems are difficult to solve. We thus proceed to construct an initial value problem whose solution has a large time limit that satisfies the given boundary value problem by making use of the occurrence of the variable t in the above formulation. Consider the system of equations

$$-\partial_t \underset{\sim}{x} = \{E|L\}^T_{\underset{\sim}{x}} - \left(\frac{\partial L}{\partial \underset{\sim}{\alpha}} \underset{\sim}{B}^{-1} \underset{\sim}{A} \right)^T \overset{\text{def}}{\equiv} \underset{\sim}{w} \quad (T = \text{transpose}) \tag{5.11}$$

$$\partial_t \underset{\sim}{\alpha} = - \underset{\sim}{B}^{-1} \underset{\sim}{A} \, \partial_t \underset{\sim}{x} = - \underset{\sim}{B}^{-1} \underset{\sim}{A} \underset{\sim}{w} \tag{5.12}$$

$$\underset{\sim}{x}(a,t) = \gamma_1 , \quad \underset{\sim}{x}(b,t) = \gamma_2 , \tag{5.13}$$

$$\underset{\sim}{x}(z,0) = \underset{\sim}{x}_o(z) , \quad \underset{\sim}{x}_o(a) = \underset{\sim}{\gamma}_1 , \quad \underset{\sim}{x}_o(b) = \underset{\sim}{\gamma}_2 . \tag{5.14}$$

Use the procedure given in Section 3 and the fact that $\det(\underset{\sim}{B}) \neq 0$ allows us to obtain the compatible initial data

$$\underset{\sim}{\alpha}(z,0) = \underset{\sim}{\alpha}_o(z) \tag{5.15}$$

such that

$$\underset{\sim}{f}(\underset{\sim}{x}_o(z), \underset{\sim}{\alpha}_o(z)) = \underset{\sim}{0} \tag{5.16}$$

for all z in $[a,b]$. It now remains to show that this initial value problem will have solutions whose large time limits satisfy the given implicit boundary value problem.

It follows immediately from (5.8) with $\delta \equiv \partial_t$ that

$$\frac{d}{dt} L(\underset{\sim}{x};\underset{\sim}{\alpha}) = \int_a^b \left\{ \{E|L\}_{\underset{\sim}{x}} - \frac{\partial L}{\partial \underset{\sim}{\alpha}} \underset{\sim}{B}^{-1} \underset{\sim}{A} \right\} \partial_t \underset{\sim}{x} \, dz ,$$

and hence (5.11) yields

$$\frac{d}{dt} L(\underset{\sim}{x};\underset{\sim}{\alpha}) = - \int_a^b \underset{\sim}{w}\, \underset{\sim}{w}^T \, dz \leq 0 \tag{5.17}$$

with equality holding if and only if $\underset{\sim}{w} = 0$; that is, if and only if (5.8) holds, in which case (5.12) yields $\partial_t \underset{\sim}{\alpha} = \underset{\sim}{0}$. Now, we have $L(\underset{\sim}{x},\underset{\sim}{\alpha}) \geq J$, $\frac{d}{dt} L(\underset{\sim}{x},\underset{\sim}{\alpha}) \leq 0$ and hence we obtain a contradiction unless $\lim_{t\to\infty} \frac{d}{dt} L(\underset{\sim}{x},\underset{\sim}{\alpha}) = 0$. We have seen, however, that this can be the case if and only if $\lim_{t\to\infty} \underset{\sim}{w} = \underset{\sim}{0}$, in which case we also obtain $\lim_{t\to\infty} \partial_t \underset{\sim}{\alpha} = \underset{\sim}{0}$. The desired result is then established on noting that all solutions of (5.12) satisfy $\underset{\sim}{f}(\underset{\sim}{x},\underset{\sim}{\alpha}) = \underset{\sim}{0}$ since $\det(\underset{\sim}{B}) \neq 0$.

REFERENCES

1. Davidenko, D.F. : *Ukr. Mat.* Z.5, 196 (1953): Branin, F.H., *Memoirs IEEE Conference on Systems, Networks and Computers* (Oaxtepec, Mexico, 1971).

2. Gelfand, I.M. and S.V. Fomin, *Calculus of Variations* (Prentice-Hall, Englewood Cliffs, 1963).

Numerical Solution of Large Systems of Stiff Ordinary Differential Equations in a Modular Simulation Framework

A. I. Johnson, and J. R. Barney

1. INTRODUCTION

A stiff ordinary differential equation (o.d.e.) is one in which one component of the solution decays much faster than others. Many chemical engineering systems give rise to systems of stiff o.d.e.s. Such situations occur in chemical reactors where the rate constants for the reactions involved are widely separated, and in multistage systems, for example in a distillation column, because the time constants for the various components and plates may be very different and also since the time constant for the reboiler is large compared with the time constant of a plate.

2. NUMERICAL SOLUTION OF STIFF O.D.E.S.

Most realistic stiff systems do not have analytical solutions so that a numerical procedure must be used. Conventional methods such as Euler, explicit Runge-Kutta and Adams-Moulton are restricted to a very small step size in order that the solution be stable. This means that a great deal of computer time could be required.

A common approach is to simply eliminate those differential equations having small time constants and to solve them as algebraic equations instead. This technique, known as the pseudo steady state approach, is very useful in those situations where it can be applied. However, it is not generally applicable and it must be applied with great care to ensure a reasonably accurate solution.

In the past decade, many new sophisticated methods have been developed to overcome the instability problem. Major literature surveys of these methods appear in Bjurel *et al.* (1970) and Willoughby (1974). These methods comprise a wide variety of both explicit and implicit methods.

For an implicit method, repeated iteration converges for only a very small step size so that a method such as Newton-Raphson must be used. Each Newton-Raphson iteration requires the solution of a system of linear algebraic equations equal in size to the number of o.d.e.s. The matrix of coefficients is a function of the system Jacobian matrix and is usually quite sparse. Considerable computer time can thus be required to solve a large stiff system unless special sparse techniques are employed. The following section describes testing of various linear equation solvers for this purpose.

3. NUMERICAL SOLUTION OF SPARSE LINEAR ALGEBRAIC EQUATIONS

Several conventional methods and some sparse techniques were tested on a set of banded test examples. We will assume that our "sparse" equations contain at most 15 nonzeros per equation.

3.1 Methods Tested For Solving Linear Algebraic Equations

Five conventional methods were studied:

(1) MINV An SSP matrix inversion algorithm (SSP, 1970)
(2) SOLVE Gaussian elimination with no pivoting
(3) DECOMP-SOLVE Gaussian elimination with partial pivoting (Forsythe and Moler, 1967)
(4) JACOBI Jacobi iterative method (initial estimate of zero used)
(5) GAUSS-SEIDEL Gauss-Seidel iterative method (initial estimate of zero used)

Several methods oriented to sparse systems were compared.

Key (1973) has written a program called SIMULT based on Gauss-Jordan elimination with seven pivoting options, all of which were tested.

IMP (Brandon, 1972) is a software system developed by D. M. Brandon at the University of Connecticut. It is now offered as a standard CDC software product through the Application Services Division of CDC. An object copy of IMP, Version 1.1, FTN version 4.0 compiler and optimization level 2 was used.

Five options with IMP package were tested:

(1) Gauss-Seidel
(2) Gradient
(3) Crout elimination, fixed order, written for general sparse matrices
(4) Crout elimination, variable order, written for general sparse matrices
(5) Crout elimination, variable order, written for variable or constant banded matrices.

In addition, a program supplied by D. M. Brandon (Brandon, 1974a), which is a simplified version of the method (4) above, was used. It should run faster because there is none of the overhead associated with the IMP routines.

Schappelle (1967) developed a program called LINEQ4 which is available from VIM, the CDC 6000 Series User Association.

Bending and Hutchison (1973) developed the concept of an "operator list" to store the solution process by Gaussian elimination. Details are presented in the thesis by Barney (1975).

3.2 Numerical Testing

Banded systems of 50, 100 and 200 linear equations with bandwidths 3, 5, 7, 9, 11 and 15 were used as test systems. The nonzero elements were generated randomly by rows from -10 to +10 using the FTN library random number generator RANF (Control Data, 1971). An initial seed of one was used. The right hand side was adjusted to yield a vector of ones as the solution.

The banded systems were used mainly for convenience; none of the algorithms tested takes advantage of this structure. However, with banded matrices, no new elements will be generated during the solution; thus the testing does not reflect how well the methods handle this problem. The Bending-Hutchison method, however, does handle this easily.

Execution times for the various methods are shown in Tables 2,3 and 4. Table 1 provides a key to the methods. Details are given in Barney (1975).

The conventional methods are obviously unsuitable for these problems. Matrix inversion takes much time and Gaussian elimination is quite slow. The iterative methods, Jacobi iteration and Gauss-Seidel iteration both diverged from initial estimates of zero. None of the conventional techniques were able to solve 200 equations as this required more than the 49K of core available.

Of the sparse routines, many gave inaccurate answers. This

may be attributed to the fact that the nonzero elements were generated over a wide range, producing badly conditioned systems. If instead, the elements are generated from 1-2, these programs work quite well. The execution times were independent of the spread of the nonzero elements.

Of the seven pivoting options given by Key, the best was minimum-row, minimum-column which Key also recommends.

The Bending-Hutchison routine, TRGB2, gave the best performance of the sparse methods. This is almost to be expected since all of the work of determining the solution process has already been done by TRGB. The method is ideal for repeated solution of linear equations.

There is another program developed at IBM by Gustavson, Liniger and Willoughby (1970) in which the first run generates the FORTRAN code to solve the specific linear system. Further solutions of the same system structure can use the FORTRAN code already producted (Calahan, 1968a). This process has been implemented along with a modified version of Gear's method (Brayton, Gustavson and Hachtel (1972); Hachtel, Brayton and Gustavson, 1971) into the ECAP II Package (Branin *et al.*, 1971).

The version of TRGB-TRGB2 used in the testing stored the operator list completely in core. For 200 equations, the method could not store the list for bandwidth of 11 or more. The list can be put into auxiliary storage, however, but this will require more computer time to transfer the list or parts of it to and from core. If this is done, there would be no limit to the number of equations which can be solved. Table 5 gives the length of the operator list for the test systems.

As the bandwidth increases, one of the IMP options, method 17 of Table 1 becomes more competitive. Possibly as the bandwidth increases still further, the IMP method will become even faster than TRGB2. This opinion is held by Brandon (Brandon, 1974b).

4. NUMERICAL TESTING OF STIFF TECHNIQUES

4.1 Testing by Other Workers

Several numerical comparisons of methods for solving stiff o.d.e.s. have been made.

Steeper (1970) compares Sloate's method (Sloate, 1970), Liou's method (Liou, 1966), Pope's method (Pope, 1963), second and fourth order Runge-Kutta, Jung's variation of Treanor's method (Jung, 1967), a linear extrapolation method from the FACE program (Price, 1970) and MINTA5, a modified version of Adams-Moulton-Shell (Shell, 1958). Based on two test examples, he recommends MINTA5.

Lapidus and Seinfeld (1971), also Seinfeld, Lapidus and Hwang (1970), have compared fourth order Runge-Kutta, fourth order Adams, Treanor's method (Treanor, 1966), the modified mid-point rule, the trapezoidal rule with and without extrapolation, Calahan's method (Calahan, 1968b) and two methods of Liniger and Willoughby (1970). They found the last four implicit methods to be superior and roughly comparable in terms of accuracy and computing time. For highly stiff systems, they recommend one of Liniger and Willoughby's techniques. Four test examples were used.

Brandon (1972, 1974c) compares his own method with the stiff techniques compared by Lapidus and Seinfeld and finds his method to be superior. He uses two test examples.

Hronsky and Martens (1973) compare Runge-Kutta-Newton (Kreyszig, 1967) which is the pseudo steady state approach used with the fourth order Runge-Kutta method, Treanor's method (Treanor, 1966), a method of Liniger and Willoughby (1970), and the backward differentiation formula (Brayton *et al.*, 1972) which is a modification of Gear's method (Appendix B). Based on five test examples, they recommend the methods in the order above, except if the number of equations is less than six, Liniger and Willoughby's method is recommended over that of Treanor.

Hull and Enright have been doing extensive comparisons of stiff techniques, but their final results are not yet available. Their preliminary testing indicated the three best methods to be trapezoidal rule with extrapolation, Gear's method and the second derivative multistep method (Enright, 1974a). Enright (1974b) shows the last two methods to be comparable.

Several other comparisons appear in the literature but, unfortunately, most authors of new techniques compare their algorithm only with an inefficient conventional method such as fourth order Runge-Kutta or with a stiff technique very similar to theirs, to demonstrate a superior modification.

4.2 Test Methods and Examples

The following methods were selected for numerical testing: four conventional methods none of which are oriented to solving stiff systems:
(1) Euler
(2) Adams-Moulton-Shell (Shell, 1958)

(3) Runge-Kutta-Merson (Merson, 1957)
(4) Gear's nonstiff option (Gear, 1971a,b)

four explicit stiff techniques:
(1) Richards, Lanning and Torrey (1965)
(2) Nigro (1969(
(3) Treanor (1966)
(4) Fowler and Warten (1967)

four implicit stiff techniques:
(1) Klopfenstein and Davis (1971)
(2) Sandberg and Schichman (1968)
(3) Brandon (1972, 1974c)
(4) Gear's stiff option (Gear 1971a,b).

The IMP package containing Brandon's method was also tested. Obviously it is impractical to test all of the many methods developed, but these provide a cross-section of promising explicit and implicit techniques. The explicit methods have the advantage of avoiding the solution of linear algebraic equations.

The four implicit methods used matrix inversion (MINV) in solving the corrector. Gear's method was also used with two other linear equation solvers: DECOMP-SOLVE and TRGB-TRGB2. TRGB was used only on the first time step to set up the operator list and TRGB2 used thereafter.

Eleven test systems were used (Appendix A). They can be divided into three groups:
(1) small stiff systems: I, II, III, IV, V
(2) large stiff systems: VI, VII
(3) nonstiff systems: VIII, IX, X, XI

Each of the fifteen test programs were tested on each example. Ideally the same method could handle both nonstiff and stiff methods efficiently; otherwise, two methods would have to be used.

The two complex eigenvalue systems III and VIII had small imaginary parts to avoid introducing the problem of simulating rapidly oscillating components.

Most of the methods used had a local truncation error estimate: Gear, IMP, Brandon, Fowler-Warton, Klopfenstein-Davis, Sandberg-Shichman, Adams-Moulton-Shell and Merson. Gear's error control is described in Gear (1971a,b). IMP uses a complex algorithm (Brandon, 1972, 1974c) to calculate the step size. In all of these methods, however, the maximum relative local error was kept below an upper tolerance and usually above a lower tolerance (except for Gear and IMP). The upper tolerance was 10^{-3} for systems I-X and 10^{-5} for system XI. The lower tolerance was taken as one-tenth of the upper tolerance. Relative errors were used with the absolute error estimate divided by the maximum value of the corresponding independent variable. This maximum value had a minimum of 1.0, so that in effect, relative errors were used for large y values and absolute errors for small y values. Fowler-Warten had a complex error control scheme, but it was not used here. The last six methods mentioned above used the double-halving approach: doubling the step if the error is too small and halving the step if the error is too large.

Three methods had no error estimate: Euler, Nigro and Treanor. Lapidus and Seinfeld (1971) recommend the following error control scheme for Treanor's method. If the relative change in y, $\left| y_{n+1} - y_n \right| / \left| y_{n+1} \right|$, is greater than some upper tolerance, the step size is halved and the step repeated. If it is smaller than some lower tolerance, the step is doubled. Tolerances of 0.05 and 0.005 were used. This strategy gave unstable results for systems V, VI and VII. It was altered for them to keeping

$$0.001 < \left| \frac{y_{n+1} - y_n}{y_{n+1} + R} \right| < 0.01$$

where $R = 0.1$ if $y_{n+1} < 0.5$

$= 0.0$ otherwise

Euler and Nigro used a fixed step size. The critical step size was used for the stiff examples, i.e., the step size just before the onset of instability. For the nonstiff examples, the step size was adjusted to keep the final error below 1%.

For Nigro's method, the parameters corresponding to $h\lambda = 120$ were used. Euler's method was used to start the integration. A step size of 0.01 of the Nigro step size was used to calculate the starting points. This was not included in the execution time.

The first order Klopfenstein-Davis algorithm tested better than the second order. Tests with various coefficients of α, β, u, v and a yielded very similar results so the values $\alpha = 1.0$, $\beta = 0.0$, u = 0.25, v = 0.75 and a = 0.6667 were used (Klopfenstein and Davis, 1971).

4.3 Test Results

Global errors are given at the end of the simulation with

the exception of systems II, V, VI and VII. There, the error is given after 10% of the simulation, since it was expected several methods could require near infinite computer time to reach the end of the integration.

Analytical solutions are available on systems I, II, III, IV, VIII and IX to calculate the error. For the remaining examples, V, VI, VII, X and XI, the Euler method with $h = 0.00005$ was used to get an accurate solution. This was arbitrarily taken as the "correct" solution. In each case with $h = 0.0001$, the results differed at worst in the fourth significant figure.

None of the examples came close to steady state, so an error taken at the end of the simulation will be reasonably indicative of the accuracy of the trajectory.

Table 6 shows the execution times and errors for the explicit methods and Euler step size. Tables 7 and 8 show results for the explicit and implicit stiff methods respectively. Nigro's step size is also shown. Table 9 shows IMP and three different linear equation solvers in conjunction with Gear's method. Gear-MINV is repeated from Table 8.

In general the conventional methods did not perform well on the stiff examples. As expected, Euler's method was the least inefficient, but obviously a method designed specifically for stiff systems is to be preferred. Some of the runs especially on systems II, V, VI and VII were a waste of computer time, as the methods were so inefficient. The conventional methods are not of great concern here except to note how well the stiff methods were able to solve the nonstiff examples by comparison. They do reasonably well with respect to execution time and error. Since the execution times are so small anyway, and reasonable accuracy is attained, it would suffice to use a stiff method for both nonstiff and stiff examples.

Hull *et al.* (1972a) consider a method of Krogh (1969) very good for nonstiff systems, but they rate Gear's (nonstiff option) closely behind it. In our limited testing of nonstiff examples here, Runge-Kutta-Merson did very well. AMOS could not handle system XI, probably because of the extremely small tolerance demanded.

In general, the explicit stiff methods did not perform as well as the implicit methods. Of the stiff methods, Gear's was easily the best (Table 8). The execution times were small and the errors acceptable. In system X, one of the extremely small components ($\sim 10^{-5}$) had a large relative error at the end of the simulation. This is not serious since the variable is almost zero, unless the user wants an accurate value for it. The second largest error is given in brackets corresponding to a much larger component. Gear's method can give large relative errors for extremely small components.

The Fowler-Warten method handled the complex eigenvalue systems III and VIII reasonably well in spite of the fact that it is intended for systems with the largest eigenvalue real.

Neither of the step-changing strategies used with Treanor's method were very efficient. Possibly a fixed step size would be a better strategy.

Some of the methods simulated only the extremely small components inaccurately especially in Systems V and X. In these cases, the second largest error is given in brackets.

Different methods of handling the corrector are illustrated in Table 9 with Gear's method. For less than about five equations, matrix inversion is fastest, then Gaussian elimination is preferable up to perhaps ten equations, but for large systems a sparse method such as the Bending-Hutchison algorithm is necessary, as shown by the results for systems VI and VII. The difference in execution time is slight for the small examples.

System XI ran slightly faster with the tridiagonal option (0.753 seconds) rather than with TRGB-TRGB2 (0.980 seconds). The time savings will increase with the size of the system.

If we had to choose one method to handle all systems stiff and nonstiff, large and small, it would be Gear's method with stiff option and the Bending-Hutchison method to solve the corrector. It handles small stiff systems and nonstiff systems efficiently enough since little computer time is required anyway. However, the nonstiff option is also available in the same program and only a switch is required to change from one to the other, so that both stiff and nonstiff options are conveniently available. Different coefficients are used, but most of the code is similar for both options.

5. THE DYNSYS 2.0 EXECUTIVE PROGRAM

Gear's method was found to be very effective in solving stiff systems. The Bending-Hutchison approach was found particularly appropriate for solving large systems of linear algebraic equations which arise in using an implicit stiff method. These new techniques were implemented into the DYNSYS package (Johnson, 1971), a modular dynamic simulation executive program. The resulting program is called DYNSYS 2.0 (Barney, Johnson and Pulido, 1975).

Both the nonstiff and stiff options of Gear's method are available. In the modular approach, each piece of equipment in a plant simulation is usually represented by a corresponding module or FORTRAN subroutine within DYNSYS. For a stiff module, the Jacobian matrix must be supplied and Newton-Raphson iteration of the corrector is used. For a nonstiff module, the Jacobian matrix is not required, as either the nonstiff option will be used (if the entire plant is nonstiff) or direct iteration of the corrector will be used with the stiff option (if some other part of the plant is stiff). A special option for tridiagonal Jacobian matrices has been included.

DYNSYS 2.0 can be used to simulate large stiff systems. The present version stores the operator list and all other variables in core so that in a 50K machine, it is limited to about 200-400 equations depending on the degree of sparseness of the Jacobian. A version of DYNSYS which keeps the operator list in auxiliary storage will be created. Segments of the operator list could be stored or read in, thus allowing much larger systems to be simulated. Thousands of o.d.e.s. could then be solved, but the program would not be as fast because of the extra time required to store and access the operator lists. It would also make the program more machine dependent.

DYNSYS 2.0 has been applied to the simulation of a commercial methylamines plant and a fictitious chemical plant described by Williams (1961).

6. EPILOGUE

The numerical comparison of methods is a never-ending activity. New methods are constantly being devised and established methods improved. Gear (1975) claims to have improved his algorithm greatly while Brandon (1975) reports his method to be much faster than earlier versions because of new error analysis. He has run our eleven test examples using the latest version of IMP (2.0) and his results are shown in Table 10. The times are faster than the IMP 1.1 we used and in some cases faster than Gear's method. However, we have not yet compared Gear's latest version with IMP. Since testing of algorithms could easily go on forever, we have decided to stop here. However, more testing will be done in the future, especially between Gear's and Brandon's methods and the evolution of DYNSYS will include the incorporation of more efficient numerical techniques.

APPENDIX A: TEST EXAMPLES

Seven stiff examples were tested:

A.1 SYSTEM I: Moderately Stiff Linear Real Eigenvalue Example

We use the 2-dimensional linear system:

$$\dot{\underset{\sim}{y}} = \underset{\approx}{A}\,\underset{\sim}{y} = \begin{bmatrix} -a & b \\ b & -a \end{bmatrix} \underset{\sim}{y} \qquad \underset{\sim}{y}(0) = \begin{bmatrix} 0 \\ 2 \end{bmatrix} \qquad (A.1.1)$$

with analytical solution:

$$y_1 = e^{-(a-b)t} - e^{-(a+b)t}$$
$$y_2 = e^{-(a-b)t} + e^{-(a+b)t} \qquad (A.1.2)$$

We choose $a = 500.5$, $b = 499.5$ for moderate stiffness, with $y_1(0) = 0$ and $y_2(0) = 2$.

A.2 SYSTEM II: Very Stiff Linear Real Eigenvalue Example

As in SYSTEM I we choose $a = 500000.5$, $b = 499999.5$ to yield the very stiff example:

$$\begin{aligned} \dot{y}_1 &= -500000.5y_1 + 499999.5y_2 \qquad y_1(0) = 0 \\ \dot{y}_2 &= 499999.5y_1 - 500000.5y_2 \qquad y_2(0) = 2 \end{aligned} \qquad (A.2.1)$$

The stiffness ratio of 10^6 is as large as has been found in practice.

A.3 SYSTEM III: Linear Complex Eigenvalue Example

The system:

$$\begin{aligned} \dot{y}_1 &= -Ay_1 + By_2 && + (A - B - 1)e^{-t} \quad y_1(0)=2 \\ \dot{y}_2 &= -By_2 - Ay_2 && + (A + B - 1)e^{-t} \quad y_2(0)=2 \\ \dot{y}_3 &= \qquad -Cy_3 + Dy_4 && + (C - D - 1)e^{-t} \quad y_3(0)=2 \\ \dot{y}_4 &= \qquad -Dy_3 - Cy_4 && + (C + D - 1)e^{-t} \quad y_4(0)=2 \end{aligned} \qquad (A.3.1)$$

is used with analytical solution:

$$\begin{aligned} y_1 &= e^{-At}(\cos Bt + \sin Bt) + e^{-t} \\ y_2 &= e^{-At}(\cos Bt - \sin Bt) + e^{-t} \\ y_3 &= e^{-Ct}(\cos Dt + \sin Dt) + e^{-t} \\ y_4 &= e^{-Ct}(\cos Dt - \sin Dt) + e^{-t} \end{aligned} \qquad (A.3.2)$$

We choose $A = 1000$, $B = C = D = 1$ to yield a moderately stiff example.

A.4 SYSTEM IV: Krogh's Example

A nonlinear example suggested by Krogh was taken from Gear (1971b). The system:

$$\dot{z}_i = -\beta_i z_i + z_i^2 \qquad i = 1,2,3,4 \qquad z_i(0) = -1 \qquad (A.4.1)$$

has analytical solution:

$$z_i = \frac{\beta_i}{1 + c_i e^{\beta_i t}} \qquad c_i = -(1 + \beta_i) \qquad (A.4.2)$$

Use $\beta_1 = 1000$, $\beta_2 = 800$, $\beta_3 = -10$, $\beta_4 = 0.001$

A.5 SYSTEM V: Chemistry Example

An example which arose in a chemistry problem is taken from Gear (1969):

$$\begin{aligned}
\dot{y}_1 &= -0.013\, y_1 - 1000\, y_1 y_3 & y_1(0) &= 1 \\
\dot{y}_2 &= -2500\, y_2 y_3 & y_2(0) &= 1 \qquad (A.5.1) \\
\dot{y}_3 &= -0.013\, y_1 - 1000\, y_1 y_3 & & \\
&\quad -2500\, y_2 y_3 & y_3(0) &= 0
\end{aligned}$$

The range of integration is [0,50].

A.6 SYSTEM VI: Polymer Example

A large stiff system of 33 equations representing a polymerization reaction was taken from Hull *et al.* (1972b):

$$\begin{aligned}
\dot{y}_1 &= k_1 y_{12}^2 - k_2 y_1 y_{23} + \frac{2D}{h^2}\left[y_2 - \left(1 + \frac{0.01h}{D}\right) y_1\right] \\
\dot{y}_i &= k_1 y_{11+i}^2 - k_2 y_i y_{22+i} + \frac{D}{h^2}\left(y_{i-1} - 2y_i + y_{i+1}\right) \quad i=2,3,\ldots,10 \\
\dot{y}_{11} &= k_1 y_{22}^2 - k_2 y_{11} y_{33} + \frac{2D}{h^2}\left(y_{10} - y_{11}\right) \qquad (A.6.1) \\
\dot{y}_i &= 2k_1 y_i^2 + 2k_2 y_{i-11} y_{i+11} + k_3 y_{i+11} \quad i=12,13,\ldots,22 \\
\dot{y}_i &= 2k_1 y_{i-11}^2 - 2k_3 y_{i-22} y_i - 2k_3 y_i \quad i=23,24,\ldots,33
\end{aligned}$$

where

$$\begin{aligned}
k_1 &= 0.70 \\
k_2 &= 1.35 \\
k_3 &= 3 \times 10^{-8} \qquad (A.6.2) \\
D &= 1.14 \times 10^{-4} \\
h &= 0.00127
\end{aligned}$$

initial conditions $y_i(0) = 0 \quad i = 1,\ldots,11$
$= 9.0 \quad i = 12,1\ ,\ldots,22$
$= 0.0 \quad i = 23,24,\ldots,33$

The range of integration was [0,100].

A.7 SYSTEM VII: Stable Polymer Example

The polymer example in 2.2.5 is an unstable system. If k_2 is changed to k_1 in equations for $i = 12$-22, the system becomes a stable fictitious one. The system still has some positive eigenvalues but it is much less unstable than System VI.

A.8 SYSTEM VIII: Linear Nonstiff Complex Eigenvalue Example

The system:

$$\begin{aligned} \dot{y}_1 &= -Ay_1 + By_2 & y_1(0) &= 1 \\ \dot{y}_2 &= -By_1 - Ay_2 & y_2(0) &= 1 \\ \dot{y}_3 &= \qquad\qquad - Cy_3 + Dy_4 & y_3(0) &= 1 \\ \dot{y}_4 &= \qquad\qquad - Dy_3 - Cy_4 & y_4(0) &= 1 \end{aligned} \tag{A.8.1}$$

is used with analytical solution:

$$\begin{aligned} y_1 &= e^{-At}(\cos Bt + \sin Bt) \\ y_2 &= e^{-At}(\cos Bt - \sin Bt) \\ y_3 &= e^{-Ct}(\cos Dt + \sin Dt) \\ y_4 &= e^{-Ct}(\cos Dt - \sin Dt) \end{aligned} \tag{A.8.2}$$

A.9 SYSTEM IX: Krogh's Nonstiff Example

The equations of SYSTEM IV: Krogh's example are used but with $\beta_1 = 0.2$, $\beta_2 = 0.2$, $\beta_3 = 0.3$, $\beta_4 = 0.4$.

A.10 SYSTEM X: Nonlinear Reaction Example

A system representing a nonlinear reaction was taken from Hull *et al.* (1972a):

$$\begin{aligned} \dot{y}_1 &= -y_1 & y_1(0) &= 1 \\ \dot{y}_2 &= y_1 - y_2^2 & y_2(0) &= 0 \\ \dot{y}_3 &= y_2^2 & y_3(0) &= 0 \end{aligned} \tag{A.10.1}$$

The integration range was [0,10].

A.11 SYSTEM XI: Gas Absorber Example

A nonlinear system representing the dynamics of a gas absorber was taken from Lapidus and Seinfeld (1971):

$$\dot{y}_1 = \{-[40.8+66.7(M_1+0.08y_1)]y_1 + 66.7(M_2+0.08y_2)y_2\}/z_1+40.8v_1/z_1$$

$$\dot{y}_i = \{40.8y_{i-1}-[40.8+66.7(M_i+0.08y_i)]y_i + 66.7(M_{i+1}+0.08y_{i+1})y_{i+1}\}/z_i \quad i=2,3,4,5 \tag{A.11.1}$$

$$\dot{y}_6 = \{40.8y_5-[40.8+66.7(M_6+0.08y_6)y_6\}/z_6 + 66.7(M_7+0.08v_2)v_2/z_6$$

where

$$v_1 = v_2 = 0$$

$$z_i = (M_i + 0.16y_i) + 75 \quad i=1,2,\ldots,6 \tag{A.11.2}$$

APPENDIX B: GEAR'S METHOD

Gear's method (1971a,b) is a variable-order, variable-step, linear, predictor-corrector algorithm. There are options for nonstiff and stiff equations.

The nonstiff option uses an Adams-Bashforth predictor and an Adams-Moulton corrector.

$$\text{Predictor:} \quad y_{n+1} = y_n + h \sum_{i=1}^{k} \beta_i \dot{y}_{n+1-i} \tag{D.1}$$

$$\text{Corrector:} \quad y_{n+1} = y_n + h \sum_{i=0}^{k} \beta_i \dot{y}_{n+1-i} \tag{D.2}$$

The order may vary from one to seven.

The stiff option uses a predictor and corrector based on the backward or numerical differentiation methods.

$$\text{Predictor:} \quad y_{n+1} = h\eta_1\dot{y}_n + \sum_{i+1}^{k} \alpha_i y_{n+1-i} \tag{D.3}$$

$$\text{Corrector:} \quad y_{n+1} = h\eta_0^*\dot{y}_{n+1} + \sum_{i=1}^{k} \alpha_i^* y_{n+1-i} \tag{D.4}$$

From first to sixth order is available.

The corrector equation has the property of stiff stability for up to sixth order.

REFERENCES

Barney, J.R., Ph.D. Thesis "Dynamic Simulation of Large Stiff Systems" McMaster University, Hamilton, Canada, 1975

Barney, J.R., Johnson, A.I. and Pulido, J. 1975: Dyanmic Simulation and Control of Chemical Processes Using a Modular Approach, paper presented at Symposium on Computers in the Design and Erection of Chemical Plants, Karlovy Vary, Czechoslovakia, Aug. 31-Sept. 4

Bending, M.J. and Hutchison, H.P. 1973: The Calculation of Steady State Incompressible Flow in Large Networks of Pipes, Chem. Eng. Sci. 28, 1957

Bjurel, G., Dahlquist, G., Lindberg, B., Linde, S. and Oden, L. 1970: Survey of Stiff Ordinary Differential Equations, Rept. NA70.11, Dept. of Information Processing Computer Science, The Royal Institute of Technology, Stockholm, Sweden

Brandon, D.M. 1972: IMP - A Software System for the Direct or Iterative Solution of Large Differential and/or Algebraic Systems General Manual, Dept. of Chem. Eng., Univ. of Connecticut, Storrs, Conn.

Brandon, D.M. 1974a: private communication, March 22

Brandon, D.M. 1974b: private communication, June 24

Brandon, D.M. 1974c: A New Single-Step Implicit Integration Algorithm With A Stability and Improved Accuracy, Simulation 23,1,17

Brandon, D.M. 1975: private communication

Branin, F.H., Hogsett, G.R., Lunde, R.L. and Kugel, L.E. 1971: ECAP II - An Electronic Circuit Analysis Program, I.E.E.E. Spectrum 8,6

Brayton, R.K., Gustavson, F.G. and Hachtel, G.D. 1972: A New Efficient Algorithm for Solving Differential-Algebraic Systems Using Implicit Backward Differentiation Formulas, Proc. I.E.E.E. 60,1,98

Calahan, D.A. 1968a: Efficient Numerical Analysis of Nonlinear Circuits, Proc. Sixth Annual Allerton Conf. on Circuit and System Theory, P. 321

Calahan, D.A. 1968b: A Stable, Accurate Method of Numerical Integration for Nonlinear Systems, Proc. I.E.E.E. (Letters) 56,744

Control Data 1971: CYBER 70 Computer System Models 72, 73, 74, 76, 7600 Computer System, 6000 Computer Systems, Fortran Extended Version 4, Reference Manual, Control Data Corporation, Sunnyvale, California

Enright, W.H. 1974a: private communication

Enright, W.H. 1974b: Second Derivative Multistep Methods for Stiff Ordinary Differential Equations, SIAM J. Numer. Anal. 11,2, 321

Forsythe, G. and Moler, C.B. 1967: Computer Solution of Linear Algebraic Systems, Prentice-Hall

Fowler, M.E., and Warten, R.M. 1967: A Numerical Integration

Technique for Ordinary Differential Equations with Widely Separated Eigenvalues, I.B.M.J. Res. Dev. 11,537

Gear, C.W. 1969: The Automatic Integration of Stiff Ordinary Differential Equations, Information Processing 68 (Proc. IFIP Congress 1968), North-Holland Publishing Co., Amsterdam, P. 187

Gear, C.W. 1971a: Algorithm 407, DIFSUB for Solution of Ordinary Differential Equations, Comm. A.C.M. 14,3,185

Gear, C.W. 1971b: Numerical Initial Value Problems in Ordinary Differential Equations, Prentice-Hall

Gear, C.W. 1975: private communication

Gustavson, F.G., Liniger, W. and Willoughby, R. 1970: Symbolic Generation of an Optimal Crout Algorithm for Sparse Systems of Linear Equations, J.A.C.M.17,1,87

Hachtel, G.D., Brayton, R.K. and Gustavson, F.G. 1971: The Sparse Tableau Approach to Network Analysis and Design, I.E.E.E. Trans. Circuit Theory CT-18,1,101

Hronsky, P. and Martens, H.R. 1973: Computer Techniques for Stiff Differential Equations, Proc. Summer Simulation Conf., Montreal, July 17-19, P. 131

Hull, T.E., Enright, W.H., Fellen, B.M. and Sedgwick, A.E. 1972a: Comparing Numerical Methods for Ordinary Differential Equations, SIAM J. Numer. Anal. 9,4,603

Hull, T.E., Enright, W.H. and Sedgwick, A.E. 1972b: Memorandum on Testing of Stiff Techniques, Dept. of Computer Science, University of Toronto, May 7

Johnson, A.I. 1971: Dynamic Process Simulation, AIChE Today Series

Jung, H.P. 1967: Numerical Integration of "Stiff" Equations, Rept. TIS-67SD245, General Electric Co., Syracuse, N.Y.

Key, J.E. 1973: Computer Program for Solution of Large, Sparse, Unsymmetric Systems of Linear Equations, Int. J. Num.Meth.Engng. 6, 497

Klopfenstein, R.W. and Davis, C.B. 1971: PECE Algorithms for the Solution of Stiff Systems of Ordinary Differential Equations, Math. Comp. 25,115,457

Kreyszig, E. 1967: Advanced Engineering Mathematics, Wiley

Krogh, F.T. 1969: A Variable Step Variable Order Multistep Method for the Numberical Solution of Ordinary Differential Equations, Information Processing 68, (Proc. IFIP Congress 1968), North-Holland Publishing Co., Amsterdam, P. 194

Krogh, F.T. 1973: On Testing a Subroutine for the Numerical Integration of Ordinary Differential Equations, J.A.C.M. 20,4,545

Lapidus, L. and Seinfeld, J.H. 1971: Numerical Solution of Ordinary Differential Equations, Academic Press

Liniger, W. and Willoughby, R.A. 1970: Efficient Integration Methods for Stiff Systems of Ordinary Differential Equations, SIAM J. Numer. Anal. 7,1,47

Liou, M.L. 1966: A Novel Method of Evaluating Transient Response, Proc. I.E.E.E. 54,1,20

Merson, R.H. 1957: An Operational Method for the Study of Integration Processes, Proc. of Conference on Data Processing and Automatic Computing Machines, Weapons Research Establishment, Salisbury, Australia, June 3-8, P. 110-1

Nigro, B.J. 1969: An Investigation of Optimally Stable Numerical Integration Methods with Application to Real Time Simulation, Simulation 13,253

Pope, D.A. 1963: An Exponential Method of Numerical Integration of Ordinary Differential Equations, Comm. A.C.M. 6,8,491

Price, W.W. 1970: FACE-A Digital Dynamic Simulation Program, Rept. DF-70-EU-2072, General Electric Co., Syracuse, N.Y.

Richards, P.I., Lanning, W.D. and Torrey, M.D. 1965: Numerical Integration of Large Highly-Damped, Nonlinear Systems, SIAM Rev. 7,3,376

Sandberg, I.W. and Shichman, H. 1968: Numerical Integration of Systems of Stiff Nonlinear Differential Equations, Bell Systems Technical J., 47,511

Schappelle, R. 1967: LINEQ4, A FORTRAN Subroutine for the Solution of Sparse Linear Equations, available through CDC User Association VIM, VIM F4 GDC CSLNQ4

Seinfeld, J.H., Lapidus, L. and Hwang, M. 1970: Review of Numerical Techniques for Stiff Ordinary Differential Equations, I.E.C. Fund. 9,2,266

Shell, D.L. 1958: General Electric Co., Tech. Information Ser. No. DF58AGT679, General Electric Co., Cincinnati

Sloate, H. 1970: An Implicit Formula for the Integration of Stiff Network Equations, Rept. TIS-70ELS2, General Electric Co., Syracuse, N.Y.

SSP 1970: System/360 Scientific Subroutine Package Version III, Programmers Manual, 5th Ed., GH20-0205-4, I.B.M., White Plains, N.Y.

Steeper, D.E. 1970: Numerical Integration of Stiff Differential Equations, Rept. DF-70-IE-111, General Electric Co., Syracuse,N.Y.

Treanor, C.E. 1966: A Method for the Numerical Integration of Coupled First-Order Differential Equations with Greatly Different Time Constants, Math. Comp. 20,39

Williams, T.J. 1961: Systems Engineering for the Process Industries, McGraw-Hill

Willoughby, R.A. 1974: Stiff Differential Systems, Plenum Press, New York. (Proc. International Symposium on Stiff Differential Systems, Wildbad, Germany, Oct. 4-6, 1973)

Table 1: Key To Linear Equation Solvers

NUMBER	METHOD
	CONVENTIONAL METHODS
1	(1) Matrix Inversion (MINV)
2	(2) Gaussian Elimination, No Pivoting (SOLVE)
3	(3) Gaussian Elimination, Partial Pivoting (DECOMP-SOLVE)
4	(4) Jacobi Iteration
5	(5) Gauss Seidel Iteration
	KEY
6	(1) Simple Gauss-Jordan Elimination
7	(2) Gauss-Jordan Partial Pivoting
8	(3) Gauss-Jordan Full Pivoting
9	(4) Minimum Row-Minimum Column
10	(5) Minimum Column-Minimum Row
11	(6) Maximum Column-Minimum Row
12	(7) Minimum of Row Entries Times Column Entries
	IMP
13	(1) Gauss-Seidel
14	(2) Gradient
15	(3) Crout, Fixed Order, General Sparse
16	(4) Crout, Variable Order, General Sparse
17	(5) Crout, Variable Order, Variable Or Constant Band
	MISCELLANEOUS
18	(1) Brandon, Crout, Variable Order, General Sparse
19	(2) LINEQ4
20	(3) TRGB
21	(4) TRGB2

Table 2: Execution Times for 50 Linear Equations

NO.	BANDWIDTH						
	3	5	7	9	11	13	15
1	5.17	5.15	5.13	5.15	5.18	5.20	5.16
2	0.838	0.850	0.816	0.833	0.819	0.818	0.818
3	0.790	0.791	0.780	0.779	0.755	0.764	0.756
4	DIV.	DIV.	DIV.	DIV.	DIV.	DIV.	DIV.
5	DIV.	DIV.	DIV.	DIV.	DIV.	DIV.	DIV.
6	0.647	0.798	0.935	1.09	1.27	1.44	1.62
7	0.748	0.986	1.23	1.38	1.64	1.91	2.14
8	0.523	1.79	2.30	3.27	5.05	5.39	5.15
9	0.215	0.327	0.495	0.691[I]	0.897	1.12[I]	1.36[SI]
10	0.293	0.437	0.612	0.814[I]	1.06	1.31[SI]	1.60
11	0.768	0.970	1.22	1.49	1.78	2.06	2.39
12	0.320	0.538	0.736	0.994[I]	1.28	1.58[SI]	1.88
13	DIV.	DIV.	DIV.	DIV.	DIV.	DIV.	DIV.
14	DIV.	DIV.	DIV.	DIV.	DIV.	DIV.	DIV.
15	0.112	0.161	0.243	0.373	0.537	0.771	1.02
16	0.101	0.158	0.223	0.298	0.391	0.509	0.645
17	0.047	0.060	0.080	0.092	0.111	0.131	0.138
18	0.073	0.097	0.161	0.234	0.320	0.424	0.564
19	0.960[I]	1.35[I]	1.68[I]	2.04[I]	2.46[I]	2.97[I]	3.52[I]
20	0.185	0.268	0.346	0.453	0.550	0.674	0.797
21	0.012	0.023	0.033	0.049	0.065	0.086	0.106

I - inaccurate SI - slightly inaccurate (1-3 figure accuracy)

DIV - diverged

Table 3: Execution Times for 100 Linear Equations

NO.	BANDWIDTH						
	3	5	7	9	11	13	15
1	40.1	39.8	40.0	39.8	39.7	39.8	40.6
2	6.45	6.43	6.37	6.28	6.32	6.26	6.26
3	5.81	5.78	5.72	5.67	5.62	5.59	5.54
4	DIV.	DIV.	DIV.	DIV.	DIV.	DIV.	DIV.
5	DIV.	DIV.	DIV.	DIV.	DIV.	DIV.	DIV.
6	2.60	3.15	3.71	4.32	4.94	5.57	6.20
7	2.07	3.88	4.79	5.69	6.40	7.47	8.33
8	1.91	8.68	11.6	15.1	24.7	24.9	34.9
9	0.722 I	1.05 I	1.49 I	1.95 I	2.48 I	3.10 I	3.76 I
10	1.02 I	1.38 I	1.84 I	2.35 I	2.95 I	3.65 I	4.37 I
11	3.05	3.81	4.64	5.51	6.43	7.42	8.47
12	1.22	1.81	2.51	3.25	4.08	4.97	5.86
13	DIV.	DIV.	DIV.	DIV.	DIV.	DIV.	DIV.
14	DIV.	DIV.	DIV.	DIV.	DIV.	DIV.	DIV.
15	0.179	0.286	0.463	0.741	1.10	1.58	2.20
16	0.277	0.379	0.518	0.674	0.879	1.13	1.43
17	0.118	0.144	0.178	0.216	0.249	0.295	0.327
18	0.152	0.247	0.350	0.503	0.690	0.923	1.20
19	2.36 I	3.87 I	5.03 I	6.40 I	7.95 I	9.74 I	11.7 I
20	0.651	0.933	1.24	1.54	1.88	2.25	2.67
21	0.023	0.044	0.068	0.101	0.138	0.181	0.229

I - inaccurate SI - slightly inaccurate (1-3 figure accuracy)

DIV - diverged

Table 4: Execution Times for 200 Linear Equations

NO.	BANDWIDTH						
	3	5	7	9	11	13	15
1	T.L.	T.L.	T.L.	T.L.	T.L.	T.L.	T.L.
2	T.L.	T.L.	T.L.	T.L.	T.L.	T.L.	T.L.
3	T.L.	T.L.	T.L.	T.L.	T.L.	T.L.	T.L.
4	T.L.	T.L.	T.L.	T.L.	T.L.	T.L.	T.L.
5	T.L.	T.L.	T.L.	T.L.	T.L.	T.L.	T.L.
6	10.4	12.5	14.8	17.1	19.4	21.8	24.2
7	6.17	15.5	20.0	23.3	26.4	29.9	33.8
8	6.64	33.5	60.3	95.2	110.0	147.7	148.2
9	2.60[I]	3.57[I]	4.70[I]	5.98[I]	7.43[I]	9.00[I]	10.8[I]
10	3.85[I]	4.85[I]	6.05[I]	7.51[I]	8.97[I]	10.7[I]	12.7[I]
11	12.2	15.0	18.0[I]	21.1[I]	24.4[I]	27.7[I]	31.1[I]
12	4.56[I]	6.59[I]	8.81[I]	11.2[I]	13.7[I]	16.5[I]	19.3[I]
13	DIV.	DIV.	DIV.	DIV.	DIV.	DIV.	DIV.
14	DIV.	DIV.	DIV.	DIV.	DIV.	DIV.	DIV.
15	0.293	0.530	0.897	1.46	2.22	3.24	4.52
16	0.816	1.02	1.28	1.62	2.04	2.56	3.17
17	0.315	0.372	0.440	0.506	0.577	0.650	0.743
18	0.417	0.587	0.824	1.15	1.53	2.00	2.59
19	7.26[I]	13.2[I]	17.5[I]	22.7[I]	28.5[I]	35.1[I]	42.3[I]
20	2.45	3.47	4.51	5.59	T.L.	T.L.	T.L.
21	0.046	0.086	0.137	0.201	T.L.	T.L.	T.L.

I - inaccurate SI - slightly inaccurate (1-3 figure accuracy)

T.L. - too large (requires more than 49K) DIV - diverged

Table 5: Length of TRGB Operator List

BANDWIDTH	NUMBER OF EQUATIONS		
	50	100	200
3	600	1200	2400
5	1176	2376	4776
7	1928	3928	7928
9	2848	5848	11848
11	3928	8128	-
13	5160	10760	-
15	6536	13736	-

Table 6: Execution Times and Errors For Conventional Methods

TEST SYSTEM	METHOD EULER	STEP SIZE	ERROR	AMOS	ERROR	RKM	ERROR	GEAR NON-STIFF	ERROR
I	0.006	1.9×10^{-3}	9.50×10^{-4}	0.872	1.88×10^{-3}	0.301	3.72×10^{-4}	2.30	1.13×10^{-3}
II	71.7	1.9×10^{-6}	9.51×10^{-8}	866.0	1.01×10^{-3}	300.3	1.36×10^{-4}	2188.0	3.81×10^{-4}
III	0.325	1.9×10^{-3}	4.28×10^{-3}	2.06	4.29×10^{-3}	1.18	3.55×10^{-3}	4.90	4.05×10^{-3}
IV	1.47	1.9×10^{-3}	9.62×10^{-6}	9.09	5.65×10^{-4}	6.02	7.27×10^{-6}	36.4	2.89×10^{-5}
V	17.1	5.0×10^{-4}	3.91×10^{-5}	240.0	114.6 (4.70×10^{-3})	79.0	154.6 (5.21×10^{-4})	431.6	0.368 (3.95×10^{-4})
VI	U[1] 23.0 (0-10)	2.5×10^{-4} 1.0×10^{-3}	4.21×10^{-4}	>1500 101.0 (0-10)	9.95×10^{-3}	>1000 48.0 (0-10)	1.99×10^{-4}	1925.0	1.86×10^{-4}
VII	37.6	6.0×10^{-3}	4.55×10^{-3}	353.0	0.855[2]	167.8	7.94×10^{-2}	656.0	8.80×10^{-3}
VIII	0.530	2.0×10^{-3}	9.58×10^{-3}	0.059	6.69×10^{-3}	0.021	1.89×10^{-3}	0.101	8.15×10^{-4}
IX	1.04	5.0×10^{-3}	9.48×10^{-3}	0.107	9.62×10^{-3}	0.041	1.56×10^{-2}	0.213	6.67×10^{-3}
X	0.636	2.0×10^{-3}	9.75×10^{-3}	0.059	3.48 (1.81×10^{-4})	0.022	0.394 (6.05×10^{-4})	0.164	4.39×10^{-2}
XI	1.39	2.5×10^{-3}	9.45×10^{-3}	>320		0.263	4.87×10^{-4}	1.30	2.25×10^{-4}

[1] unstable

[2] only extremely small components inaccurate

Table 7: Execution Times And Errors For Explicit Stiff Techniques

TEST SYSTEM	METHOD RLT	ERROR	NIGRO	STEP SIZE	ERROR	TREANOR	ERROR	FOWLER WARTEN	ERROR
I	0.092	4.83×10^{-3}	0.144	1.8×10^{-3}	5.35×10^{-3}	0.384	2.24×10^{-4}	0.495	2.24×10^{-3}
II	0.068	5.03×10^{-4}	26.5	1.0×10^{-5}	2.15×10^{-2}	0.393	8.90×10^{-7}	49.1	3.98×10^{-2}
III	3.32	3.81×10^{-3}	0.498	1.8×10^{-3}	6.13×10^{-3}	1.38	5.09×10^{-4}	1.48	7.49×10^{-5}
IV	0.356	5.03×10^{-3}	0.817	5.0×10^{-3}	6.88×10^{-3}	6.90	1.85×10^{-3}	0.317	1.79×10^{-2}
V	0.078	343.0 (6.58×10^{-4})	18.5	1.0×10^{-3}	1.06×10^{-2}	0.505	3.56×10^{-3}	0.624	8.42×10^{-3}
VI	36.5	8.86[2]	464.0	1.0×10^{-3}	3.35×10^{-2}	U[1] 44.5 (0-10)	4.41×10^{-2}	117.0	2.48×10^{-2}
VII	3.58	2.03[2]	46.5	1.0×10^{-2}	1.73×10^{-2}	219.0	8.24×10^{-2}	75.2	1.41×10^{-2}
VIII	0.159	8.61×10^{-2}	1.24	2.0×10^{-3}	9.72×10^{-3}	0.927	1.97×10^{-5}	0.080	2.64×10^{-2}
IX	0.317	5.45×10^{-2}	2.02	4.0×10^{-3}	1.05×10^{-2}	2.69	3.13×10^{-9}	0.077	7.07×10^{-3}
X	0.118	0.630 (2.81×10^{-2})	1.36	2.5×10^{-3}	2.64×10^{-2}	2.36	2.46×10^{-4}	0.060	1.83×10^{-3}
XI	0.744	3.00×10^{-2}	5.65	1.0×10^{-2}	1.63×10^{-2}	5.07	4.06×10^{-5}	0.666	1.86×10^{-4}

[1]unstable

[2]large components reasonably accurate

Table 8: Execution Times And Errors For Implicit Stiff Techniques

TEST SYSTEM	METHOD KLOP DAVIS	ERROR	SAND SCH.	ERROR	BRANDON	ERROR	GEAR	ERROR
I	2.27	$5.85x10^{-4}$	0.274	$1.21x10^{-2}$	0.124	$1.22x10^{-3}$	0.139	$1.07x10^{-3}$
II	>1000 385.0 (0-10)	$5.00x10^{-8}$	0.276	$7.10x10^{-4}$	0.268	$3.98x10^{-4}$	0.196	$1.01x10^{-4}$
III	7.11	$2.67x10^{-3}$	0.788	$4.64x10^{-2}$	4.24	$3.97x10^{-3}$	0.318	$3.77x10^{-3}$
IV	39.2	$5.90x10^{-6}$	1.75	$9.41x10^{-4}$	1.11	$5.66x10^{-4}$	0.505	$1.16x10^{-5}$
V	845.0	0.267 $(7.56x10^{-3})$	0.117	$1.27x10^{-4}$	0.459	$3.10x10^{-5}$	0.168	$4.57x10^{-5}$
VI	>1000 (0-10)		>1000 626.0 (0-10)	$6.83x10^{-3}$	>1500 761.0 (0-10)	$7.72x10^{-4}$	81.7	$8.47x10^{-3}$
VII	>1000 (0-10)		442.0	0.133	299.0	$1.86x10^{-2}$	42.0	$1.71x10^{-2}$
VIII	0.760	0.206	0.721	0.204	0.294	0.308 (0.023)	0.078	$3.23x10^{-3}$
IX	0.788	0.335	0.830	0.316	0.546	$9.68x10^{-3}$	0.192	$8.07x10^{-2}$
X	0.473	0.936 $(3.0x10^{-2})$	0.455	0.870 $(2.94x10^{-2})$	0.313	$8.13x10^{-3}$	0.119	1.38 $(2.74x10^{-3})$
XI	29.1	$3.78x10^{-2}$	29.1	$3.75x10^{-2}$	2.44	$1.77x10^{-4}$	1.03	$4.85x10^{-5}$

Table 9: Execution Times and Errors for Gear's Method and IMP

TEST SYSTEM	METHOD GEAR MINV	GEAR DECOMP-SOLVE	GEAR TRGB	ERROR	IMP	ERROR
I	0.139	0.148	0.162	$1.07x10^{-3}$	0.139[1]	$1.85x10^{-3}$
II	0.196	0.216	0.235	$1.01x10^{-4}$	0.373[1]	$2.84x10^{-4}$
III	0.318	0.294	0.315	$3.77x10^{-3}$	3.72[1]	$1.06x10^{-5}$
IV	0.505	0.489	0.656	$1.16x10^{-5}$	0.967[1]	$2.69x10^{-2}$
V	0.168	0.165	0.182	$4.57x10^{-5}$	27.3[2]	$3.91x10^{-5}$
VI	81.7	20.0	7.58	$8.47x10^{-3}$	>1000	
VII	42.0	12.3	6.05	$1.71x10^{-2}$	30.2[2]	0.169
VIII	0.078	0.079	0.080	$3.23x10^{-3}$	0.258[1]	0.344
IX	0.192	0.197	0.292	$8.07x10^{-2}$	0.584[2]	$2.37x10^{-3}$
X	0.119	0.128	0.130	1.38 ($2.74x10^{-3}$)	1.12[1]	$5.27x10^{-3}$
XI	1.03	0.897	0.980 0.753(TR1)	$4.85x10^{-5}$	11.5[1]	0.967

[1]Crout elimination, variable order, variable or constant banded matrices

[2]Gauss-Seidel

Table 10: Execution Times and Errors for IMP 2.0

TEST SYSTEM	EXECUTION TIME	ERROR
I	0.0521	4.08×10^{-4}
II	0.0770	4.97×10^{-4}
III	0.0718	2.33×10^{-4}
IV	0.269	4.47×10^{-4}
V	0.136	1.15×10^{-4}
VI	26.0	7.90×10^{-3}
VII	19.3	2.75×10^{-3}
VIII	0.227	1.08×10^{-2}
IX	1.49	2.41×10^{-3}
X	0.153	8.76×10^{-3}
XI	0.100	1.56×10^{-2}

FAST:[+] A Translator for the Solution of Stiff and Nonlinear Differential and Algebraic Equations

L. F. Stutzman, F. Deschard,[*] R. Morgan,[**] and T. Koup

1. INTRODUCTION

A new translator, FAST, requiring no previous computer experience by the user, has been developed for IMP, an implicit integration package shown to be "absolutely" stable on stiff equations. IMP utilizes a novel integration algorithm with control options which provide problem specificity, thereby reducing computer time. A vector elimination approach reduces core requirements on real systems of equations with sparse coefficient matrices.

FAST has been developed to eliminate the programming effort required by IMP. Models or systems of differential and algebraic equations are read in with minimal instruction coding. Standard forcing functions and an interpolation routine for experimental data are provided, and any FORTRAN programmable function can be used.

The FAST translator used with IMP has been shown to be very competitive with existing integration, problem solving, and modeling packages.

---

* Le Freppel, Rue Freppel, 83100 Toulon, France

** Exxon, Baytown, Texas

+This paper describes the features of FAST V1.0, and an experimental version, FAST V1.1. Since then, the package has been renamed GEMS, General Equation Modeling System. The current version is GEMS V2.0 and contains all of the features of FAST V1.0 and FAST V1.1 in final form.

2. MATHEMATICAL CONDITIONS

The FAST (11, 12) translator was developed to provide an easy and efficient way to code dynamic and steady state problems which represent real systems. FAST requires only a little experience with computers and knowledge of numerical analysis. It was designed to permit a chemical engineer or a process engineer to code his own problems.

FAST utilizes IMP (1,2,3) (Implicit Modeling Program). IMP was developed to solve large and stiff differential and algebraic systems. Since FAST is a translator which utilizes IMP, a brief description of IMP is in order.

IMP is a continuous simulation language which incorporates into a single software package an implicit integration algorithm, a variety of direct and iterative methods for solving simultaneous equations, and a vector elimination method for storing and acting upon sparse matrices. There are a number of other features, some of which will be mentioned later. FORTRAN calls are used exclusively for programming a problem in IMP.

Perhaps the key feature of IMP for solving "stiff" problems is the implicit integration algorithm. The algorithm developed - called Brandon's method (1,4) - is an implicit, single step, single variable, non-linear equation represented by:

$$\overline{X}_n = \overline{X}_{n-1} + [\bar{\bar{U}}^D \dot{\overline{X}}_n + \bar{\bar{W}}^D \dot{\overline{X}}_{n-1}] \qquad (1)$$

where

$$W_i = [-Z_i^{-1} - (e^{-Z_i} - 1)^{-1}]h$$

$$U_i = (h - W_i)$$

Z_i = a diagonal transitional matrix

$$= n \frac{\partial \dot{\overline{X}}_i}{\partial \overline{X}_i} h \sum_{j=1}^{N} a_{ij}(X_j/X_i)$$

h = integration step size

This algorithm has been found to be A-stable, operates with fixed or variable step size, is computationally fast, and is accurate. It was tested first by solving the linear equation

$$\dot{\overline{X}} = \bar{\bar{A}}\,\overline{X} \qquad (2)$$

for chosen "stiff" values of the coefficient matrix $\bar{\bar{A}}$ (all variations of 10 and .1 in a four element matrix), and comparing the results with that obtained by the analytical solution. These problems were also solved by Euler explicit and implicit method and the implicit trapezoidal method. In all cases this algorithm gave answers which were remarkably close to the analytical ones (usually to within 0.1% accuracy) whereas the other numerical methods usually were off by from 1% to several fold.

For non-linear equations,

$$\dot{\overline{X}} = \overline{F}\,(\overline{X}) \tag{3}$$

the coefficient matrix $\bar{\bar{A}}$ involves a Jacobian,

$$\bar{\bar{A}} = \frac{\partial F(\overline{X})}{\partial \overline{X}} \tag{4}$$

It is proposed that for such systems, the following equation be used

$$\dot{\overline{X}} = \bar{\bar{A}}^S\ \overline{X} + \overline{B}^S \tag{5}$$

where

$$\bar{\bar{A}}^S = \left.\frac{\partial \overline{F}(\overline{X})}{\partial \overline{X}}\right|_{\overline{Y}^S}$$

and

$$\overline{B}^S = \overline{F}(\overline{Y}^S) - \left.\frac{\partial \overline{F}(\overline{X})}{\partial \overline{X}}\right|_{\overline{Y}^S} \cdot \overline{Y}^S$$

where

$\overline{Y}^S$ is a state expansion vector.

The diagonal transition matrix Z_i in the integration algorithm are averaged between the first and last iteration since $\bar{\bar{A}}$ is changing, or,

$$Z_i = \frac{h}{2} \sum_{j=1}^{N} \left(a^o_{ij} \frac{\dot{\overline{X}}^o_j}{\dot{\overline{X}}^o_i} + a^S_{ij} \frac{\dot{\overline{X}}^S_j}{\dot{\overline{X}}^S_i} \right) \tag{6}$$

To test the integration a variety of linear and non-linear sets of mathematical systems were chosen. Some of these are found in the literature (Seinfeld (5)) and have been reported as difficult problems. These were all stiff equations containing slow and fast modes and eigen values which varied by as much as 1000. Methods which were compared to this technique were Factors (6), Euler's implicit and explicit method, Trapezoidal (Crank-Nicholson), Adams, Treavor's, Calahan (7), Liniger and Willoughby (8). On all problems when considering both time and accuracy, Brandon's method proved superior.

The implicit integration algorithm is the key to the solution of stiff systems. The key to rapid solution and efficient use of computer responses for a large number of simultaneous equations are the methods for solving such equations and handling sparse matrices. Thus, a variety of direct (found best for full or nearly full matrices) and iterative (found best for sparse matrices) methods for solving simultaneous equations are optional at the discretion of the programmer. A vector elimination method (9,1,2) is built into the system for storing information and producing the solution of the problems.

Some additional features are optional methods for ordering the matrices to improve computer execution time, automatic core allocation, error estimations with automatic step size adjustment, etc. The latest version of IMP (IMPVIR) is a "virtual" version which permits a large system in the order of 100,000 equations to be solved simultaneously. These and other features are presented in the IMP Manual (2).

The philosophy behind IMP, which has been carried over into FAST, is to represent the model of all real systems - initially expressed in differential or algebraic form - in terms of state variables. For ordinary differential or algebraic systems these are in turn represented by first order systems, or,

$$\dot{\bar{X}} = \bar{\bar{A}}\ \bar{X} + B \quad \text{linear, differential} \tag{7}$$

or

$$0 = \bar{\bar{A}}\ X + \bar{B} \quad \text{linear, algebraic} \tag{8}$$

For non-linear systems,

$$\dot{\bar{X}} = f(\bar{X}) = \bar{\bar{A}}^S\ \bar{X} + \bar{B}^S \quad \text{differential} \tag{9}$$

or

$$0 = f(\bar{X}) = \bar{\bar{A}}^S\ \bar{X} + \bar{B}^S \quad \text{algebraic} \tag{10}$$

For partial differential equations, or distributed parameter systems, the equations are first discretized by dividing the system into N cells or meshes, and as a result, the problem reduces to N 1st differential-difference equations. Consider as an example the equation:

$$\frac{\partial C}{\partial t} + V\ \frac{\partial C}{\partial Z} = kC^2 + D\ \frac{\partial^2 C}{\partial Z^2} \tag{11}$$

One scheme for formulation of the above equation in differential-difference form is:

$$\frac{dC_n}{dt} = kC_n^2 - \frac{V(C_n - C_{n-1})}{\Delta Z} + D\ \frac{(C_{n+1} - 2C_n + C_{n-1})}{(\Delta Z)^2} \tag{12}$$

where n = 1,2, - - - N cells or meshes

or, in terms of state variable:

$$\frac{dX_n}{dt} = \left(\frac{V}{\Delta Z} = \frac{D}{(\Delta Z)^2}\right) X_{n-1} + \left(- \frac{V}{\Delta Z} - \frac{2D}{(\Delta Z)^2}\right) X_n\ + \frac{D}{(\Delta Z)^2}\ X_{n+1} + k\ X_n{}^2 \tag{13}$$

If the non-linear terms, kX_n^2, is linearized by a Taylor expansion as:

$$kX_n^2 = k(X_n')^2 + \left.\frac{\partial kX_n^2}{\partial X_n}\right|_{X_n'} (X_n - X_n') \tag{14}$$

then equation (13) becomes:

$$\frac{dX_n}{dt} = \left(\frac{V}{\Delta Z} + \frac{D}{(\Delta Z)^2}\right) X_{n-1} + \left(-\frac{V}{\Delta Z} - \frac{2D}{(\Delta Z)^2} + 2\,kX_n'\right) X_n + \left(\frac{D}{(\Delta Z)^2}\right) X_{n+1} - k(X_n')^2 \tag{15}$$

Equation (15) may be expressed in terms of equation (9) as:

$$\dot{\overline{X}} = \bar{\bar{A}}^S\,\overline{X} + \overline{B}^S \tag{9}$$

where the column vectors have N rows, the Jacobian forms a tri diagonal matrix of elements $2\,kX_n' - \frac{V}{\Delta Z} - \frac{D}{(\Delta Z)^2}$, $\frac{D}{(\Delta Z)^2}$, and $\frac{V}{\Delta Z} + D/(\Delta Z)^2$ and the augmented vectors, $\overline{B}^S$ has elements of $-k(X_n')^2$ around each transition point X_n'.

In addition to the solution of "stiff" systems referred to previously, many real problems which are both large and stiff have been solved with the use of IMP. These include problems in heat transmission, kinetics, distillation, and process modeling (1,3,10). On all of these problems, IMP has been stable, efficient, and is believed to be accurate.

3. FAST TRANSLATOR

The FAST translator fulfills the objectives of IMP but permits the coding of problems in a simple manner. Upon reading in a small user written program, the FORTRAN programs required by IMP are automatically written by the translator. Appropriate matrices and vectors are dimensioned, constant terms are stored, the Jacobians are written for non-linearities, and the matrix topology is determined. The FORTRAN programs are then compiled and loaded, the matrix is filled in, and IMP is called to produce the solution.

In addition to the translation of a simple input program into the FORTRAN statements for IMP, FAST has within its own library a variety of library functions for forcing function, interpolation of data values as equation inputs, and evaluation of Jacobians by perturbation. FAST also permits any FORTRAN functions, whether in the System Library or user written.

The first step in using FAST is to set up the defined problem in state-variable form (1st order ordinary or algebraic equations) as discussed previously. The problem is then coded by writing these equations in this form and adding certain control statements. As an example, consider the problem for the mechanical system:

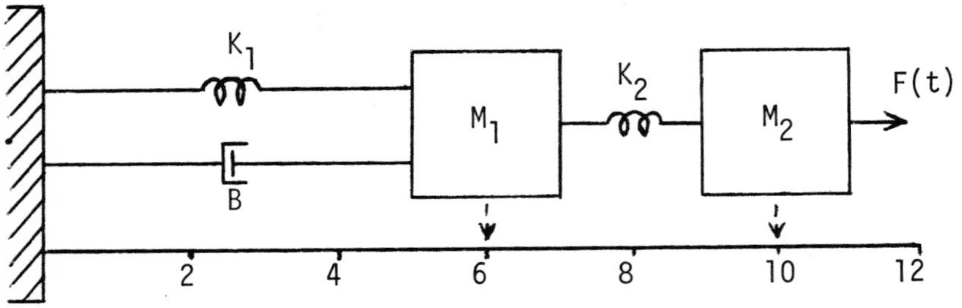

which may be represented by the following set of equations:

$$\frac{dX_0}{dX_0} = 1$$ In IMP and FAST the time variable is defined as a state variable. FAST automatically defines it to be X_0.

$$\frac{dX_1}{dX_0} = X_3 \tag{16}$$

$$\frac{dX_2}{dX_0} = X_4$$

$$\frac{dX_3}{dX_0} = \frac{K_1 + K_2}{M_1} X_1 + \frac{K_2}{M_1} X_2 - \frac{B}{M_1} X_3 + \frac{6K_1}{M_1} - \frac{4K_2}{M_1}$$

$$\frac{dX_4}{dX_0} = \frac{K_2}{M_2} X_1 + \frac{K_2}{M_1} X_2 + \frac{4K_2}{M_2} + \frac{F(X_0)}{M_2}$$

In FAST, the term $F(X_0)$ can be replaced by a forcing function. Here, a step at X0 = 1 is used.

```
FORCE, F = STEP (1.0) ;
DX1 = X3 ;
DX2 = X4 ;
DX3 = - 6 * X1 + 4 * X2 - 2 * X3 - 4.0 ;
DX4 = 2 * X1 - 2 * X2 + 8.0 + 5.0 * F (X0) ;
PARAMETERS, STEP = 0.1 ;
```

```
FINISH, X0 . GE . 5 ;
INITIAL CONDITIONS, X2 = 10., X1 = 6.0 ;
END ;
```

FAST produces diagnostics and is easy to debug since the statements are so similar to the actual mathematics.

Statements in FAST are written in free format. Each one can be coded on up to ten cards and is terminated with a semicolon. The general types of statements are as follows:

1. Comments

Cards containing an asterisk in column one are treated as comments and are ignored. They can be used to document a FAST program.

2. Model Statements

These define the model or system of equations as a set of state variables. These state variables are represented by X1, X2, X3, ... Xn. Differential equations are written as

```
DXn = expression ;
```

and algebraic equations can be written

```
Xn = expression ;
```

or

```
0 = expression ;
```

if it is not feasible to solve for Xn directly.

3. Control Statements

These convey information to FAST that can not be put in the model statements. Each begins with an identifying keyword, and the information required on the card is dependent on this keyword.

FAST uses X to represent the state variables. The programmer can define names up to 6 characters long representing constants, arrays, forcing functions, or data tables. This permits easy modification of parameters in FAST programs.

The keywords available are listed below:

(a)
```
CONSTANT, name = value, name = value, . . .;
ARRAY, name = value, value, value, . . ., name = value,
       value, value, . . . ;
COMPUTE, name = expression, name = expression, . . . ;
```

These statements assign constant numerical values to names. The names can then be used in model statements in the same manner as constants and state variables. As the ARRAY card assigns a set of values to one name, each occurrence of the name must be followed by a subscript, i.e., ARRAY, P = 1,3,5 ;

defines P(1) = 1, P(2) = 3, and P(3) = 5 ;

(b) FORCE, name = function (parameters), name = function (parameters), . . . ;

TABLE, name = X-value, X-value, . . .Y-value, Y-value . . .;

Forcing functions or interpolation of experimental data functions are requested with the above statements. The forcing functions available are a STEP, RAMP, SINE, PULSE, or PULSE TRAIN. Their form is shown in Table 1. Functions of experimental data can be supplied in a TABLE card with the independent variable points first. In referencing either type of function in a model statement, one parameter, the state variable to be used as the independent variable, is written.

(c) INITIAL, Xn = value, Xn = value, . . . ;

PARAMETER, name = value, name = value, . . . ;

Non-zero initial conditions on the state variables and any changes to the default parameters are specified as above. Many of the parameters are direct parameters to IMP, whereas others control certain aspects of FAST itself. A list of the parameters and their defaults appears in Table 2.

(d) PRINT, variable, variable, . . . ;

PREPARE, variable, variable, . . . ;

PLOT, variable = 'title'/variable = 'title', variable = 'title'/variable = 'title', . . . ;

CALCOMP PLOT, variable = 'title', variable = 'title', . . ./variable = 'title' ;

REMARK, text ;

TITLE, text ;

These statements allow the user to specify which variables are to be printed or plotted, and to supply identifying titles.

(e) FINISH, variable = expression, variable = expression, ... ;

MODIFY ;

RERUN, name, name, . . . ;

END ;

The additional features provided by these statements allow the user to control or monitor the solution directly. FINISH sets termination conditions based on any number of state variables. MODIFY indicates that a new set of cards follow, (other than COMPUTE, RERUN, or model definition equations) and the integration is to be restarted using these new data. RERUN allows the user to examine the initial and final state variables and any number of constants, and from them determine new

Table 1

FORCING FUNCTIONS

Description	Function
STEP (Xn_o) $Xn \leq Xn_o$, $f(Xn) = 0$ $Xn > Xn_o$, $f(Xn) = 1$	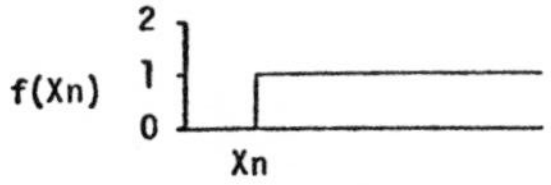
SINE (Xn_o, ω, ϕ) $Xn \leq Xn_o$, $f(Xn) = 0$ $Xn > Xn_o$, $f(Xn)$ $= \sin(2\pi\omega(Xn - t_o) + \phi)$	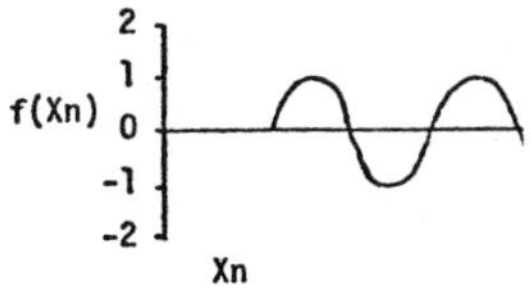
RAMP (Xn_o, m, Xn_f) $Xn \leq Xn_o$, $f(Xn) = 0$ $Xn \geq Xn_f$, $f(Xn) = f(Xn_f)$ $Xn_o < Xn < Xn_f$, $f(Xn) = (Xn - Xn_o)m$	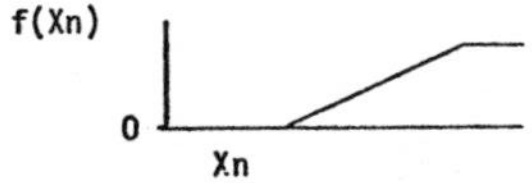
PULSE (Xn_o, w) $Xn \leq Xn_o$, $f((Xn) = 0$ $Xn \geq Xn_o + w$, $f(Xn) = 0$ $Xn_o < Xn < Xn_o + w$, $f(Xn) = 1$	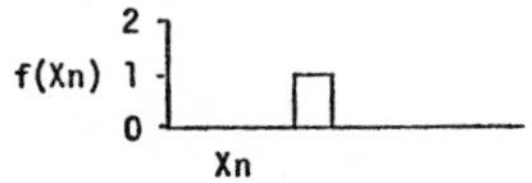
TRAIN (Xn_o, w, d, s) $s = \pm 1$ $Xn \leq Xn_o$, $f(Xn) = 0$ $m = 0, 1, 2. . .$ $Xn_o + md < Xn < Xn_o + md + w$, $f(Xn) = s^m$ $Xn_o + md + w \leq Xn \leq Xn_o + (m+1)d$, $f(Xn) = 0$	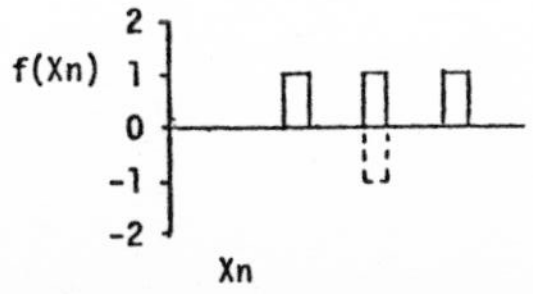

Table 2

PARAMETERS

Name	Purpose	Default
STEP	Integration step size	Minimum step size (SMIN)
SMIN	Minimum step size	10^{-4}
AUTO FIXED	Automatic step size option	AUTO
DIRECT GAUSS GRAD	Solution method	DIRECT = (BAND, RETAIN) direct solution, equations ordered once for minimum bandwidth
ITER	Maximum iterations for non-linear or iterative solutions	10 for non-linearities 20 for iterative solutions
TOLER	Tolerance criterion	.0005
DMIN	Minimum divisor	10^{-6}
FILTER	Filter factor for non-linear solutions	0
SCALE	Matrix scale factor	1
DELTA	Perturbation delta	.0005
UPDATE	Number of steps before new Jacobian is to be calculated	1
COND	Matrix conditioning	0 (none)
OUTPUT	Print and plot output increments	Step size (STEP)
LIMIT	Maximum number of steps	1000
ORDER	Interpolation order	2
ERRORS	Number of errors allowed before termination	10
SIZE	Solution matrix size	variable

constants and initial conditions for the next run. END signifies the end of the FAST program. Any FORTRAN functions can be placed after this card.

Because of the methods used internally by FAST, the deck must follow certain ordering rules. All CONSTANT, ARRAY, COMPUTE, FORCE, and TABLE cards appear first, followed by the model statements in state variable order, followed by any others required. If one or more MODIFY cards are used, however, the cards which follow each may be in any order. Obviously, the END card is last, and comments may appear anywhere before it.

4. THE NON-LINEARITIES

We saw that a problem must be written in state variable form, such as:

$$\dot{\overline{X}} = \overline{\overline{A}}\cdot\overline{X} + \overline{B} \text{ or } \dot{\overline{X}} = \overline{F}(\overline{X})$$

$$\overline{0} = \overline{\overline{A}}\cdot\overline{X} + \overline{B} \text{ or } \overline{0} = \overline{F}(\overline{X})$$

With IMP, one must calculate the Jacobian terms when the problem has some non-linearities. $\overline{F}(\overline{X})$ can be expressed with the Jacobian matrix $\overline{\overline{J}}(\overline{X})$ and the augmented constant vector $\overline{C}$, such as:

$$\overline{F}(\overline{X}) = \overline{\overline{J}}\cdot\overline{X} + \overline{C}$$

$$\text{where: } \overline{\overline{J}} = \frac{\partial\overline{F}(\overline{X})}{\partial\overline{X}} \quad \left\{\begin{matrix} \dfrac{\partial F_1(\overline{X})}{\partial X_1} & \cdots & \dfrac{\partial F_1(\overline{X})}{\partial X_n} \\ \cdots & \cdots & \cdots \\ \dfrac{\partial F_n(\overline{X})}{\partial X_1} & \cdots & \dfrac{\partial F_n(\overline{X})}{\partial X_n} \end{matrix}\right\}$$

$$\overline{C} = \overline{F}(\overline{X}) - \frac{\partial\overline{F}(\overline{X})}{\partial\overline{X}}\cdot\overline{X}$$

or, at the nth discrete integration step, as:

$$\overline{F}(\overline{X}_n) = \overline{\overline{J}}_{n_o}\cdot\overline{X}_n + \overline{C}_{n_o}$$

where n_o indicates an expansion point around which the Taylor expansion can be written,

$$\dot{\overline{X}}_n = \overline{F}(\overline{X}_n) = \overline{\overline{J}}_{n_o}\cdot\overline{X}_n + \overline{C}_{n_o}$$

$$\text{or } \dot{\overline{X}}_n = \frac{\partial\overline{F}(\overline{X}_{n_o})}{\partial\overline{X}_{n_o}}\cdot\overline{X}_n + \overline{F}(\overline{X}_{n_o}) - \frac{\partial\overline{F}(\overline{X}_{n_o})}{\partial\overline{X}_{n_o}}\cdot\overline{X}_{n_o}$$

The translator has the capability to evaluate the partial derivatives

$\dfrac{\partial \bar{F}(\bar{X}_{n_o})}{\partial \bar{X}_{n_o}}$ at the expansion point; these partials are calculated analytically unless they involve functions or interpolated experimental data, in which case they are approximated by the perturbation technique:

$$\frac{\partial F_j}{\partial X_i} = \frac{F_j(X_1, \ldots, X_i + \delta_i, \ldots X_n) - F_j(X_1, \ldots, X_i, \ldots, X_n)}{\delta_i} \quad \text{where}$$

$\delta_i = \max(|\Delta X_i, \varepsilon)$, the perturbation on variable X_i. The default value of Δ is $5.0\ 10^{-4}$, which may be changed by the user, and ε is a small number which becomes the perturbation value when the product ΔX_i is too small. It is a function of different error parameters. $\bar{\bar{J}}_{n_o}$ and $\bar{C}_o$ are calculated during each step of integration and stored in the appropriate locations in the IMP storage arrays.

When the partials can be calculated analytically, FAST can build the expressions for each element $\bar{\bar{J}}_{n_o}$ and $\bar{C}_o$ directly. The routine can recognize and derivate most expressions, including those involving exponentials, logarithms, transcendentals, and hyperbolics. If any user functions, forcing functions, or interpolated data functions occur, naturally an analytical partial cannot be generated. To save execution time during the actual solution, and to improve accuracy, as much of $\bar{\bar{J}}_{n_o}$ and $\bar{C}_o$ as possible will be filled in analytically. Only those terms containing functions need to be perturbated, and even then, as much as can be done analytically is done first. For example:

$$\frac{\partial[X_1{}^2 \text{ arbfun } (X_1)]}{\partial X_1}$$

is of the form

$$\frac{\partial[UV]}{\partial X_1}$$

where $U = X_1{}^2$

$V = \text{arbfun } (X_1)$

Analytically, one can obtain:

$$\frac{\partial[UV]}{\partial X_1} = U\frac{\partial V}{\partial X_1} + V\frac{\partial U}{\partial X_1}$$

$$= X_1^2 \frac{\partial[\mathrm{arbfun}(X_1)]}{\partial X_1} + \mathrm{arbfun}\ (X_1)(2X_1)$$

At this point, only the function requires perturbation, not the entire expression, eliminating the inaccuracies introduced by applying perturbation to the whole term. The additional time required to evaluate this partial term is offset by having a smaller expression to perturbate, where it must be evaluated twice.

Each equation is analyzed independently from the others due to the vector algebra approach of IMP. This enables one to limit the amount of storage required by the translator; on the other hand it also has some disadvantages, for example, if the identical function occurs in two different equations, the translator will perturbate it twice.

5. DESCRIPTION OF THE TRANSLATOR'S OPERATION

The FAST system has three basic phases:

(1) The program is read and translated into FORTRAN. The FORTRAN routines allocate the storage arrays, initialize the matrices and parameters, and contain the code necessary to compute $\bar{J}_n$ and $\bar{C}_o$. As all partials that can have been evaluated analytically, the FORTRAN programs produced can contain these expressions, so they need not be re-evaluated at each step.

(2) The resulting programs and any user supplied FORTRAN functions are compiled, and the object modules produced loaded along with IMP and any FORTRAN or FAST system modules required.

(3) Execution of the translated programs to produce answers.

A user of IMP would perform phases 2 and 3, that is, compilation and execution, in exactly the same manner as FAST does. It is, therefore, phase 1 which is of interest. It is here that FAST essentially writes the IMP programs and subroutines for the user from the set of model statements given.

As each model statement is scanned, or parsed, the individual terms are divided into three classifications: linear terms, containing only numerical constants and one state variable (i.e., 9.72/ 5.24 * X4), pseudo-linear terms containing user defined constants (i.e., FLOW/VOLUME * X7), and non-linear terms. First, linear terms are discarded, and the coefficient stored in the Jacobian matrix. Those terms containing references to forcing functions or interpolated data tables have those references changed to calls to the appropriate FAST library functions. Finally, terms containing no state variables are discarded and stored in the constant vector. All other terms remain unchanged.

At the same time it is parsing the equation, the translator identifies each state variable which appears, and which of those appear only in linear terms. This allows the allocation of storage for the matrix. The totally linear terms are now complete and can be disregarded, as the coefficient which was stored is the partial in each case. The partials of all other state variables must now be found and added into the matrix.

The parsed equation is now differentiated with respect to each remaining state variable. The partial derivatives of pseudo-linear terms can be found by inspection, and are done first. By repeated application of the chain rule, non-linear expressions are broken down into simple terms for which the derivatives can be easily computed, standard functions whose derivatives are stored in memory, and user, FAST, or FORTRAN functions which must be perturbated. Such functions are perturbated, however, only if the state variable with respect to which the partial is being taken appears in the function parameters. Otherwise, the partial is zero and the entire term drops out. The resulting expression is used to generate FORTRAN code to augment $\bar{J}_{n_0}$ and $\bar{C}_0$.

Each of the control statement types is parsed by a separate subroutine. The data from each are generally stored in core or on a disk file. At the end of the translation, all of this data are written to files where they are accessible during the execution of IMP. A simple flowchart showing some of the major FAST translator routines is given in Figure 1.

6. EXAMPLES

Example 1: A Biological Reactor

Consider the following system where S, V, T, are concentrations and t is time,

$$\frac{dS}{dr} = \frac{SV}{0.2 + V} - 0.4\ ST$$

$$\frac{dV}{dr} = \frac{-\ 2\ SV}{0.2 + V}$$

$$\frac{dT}{dr} = \frac{0.1\ SV}{0.2 + V}$$

let us define: X_1=S, X_2=V, X_3=T (X_0=t=time by default)

The equations can now be written as:

$$\frac{dX_1}{dX_4} = \text{FUN}\ (X_1,\ X_2) - 0.4 * X_1 * X_3$$

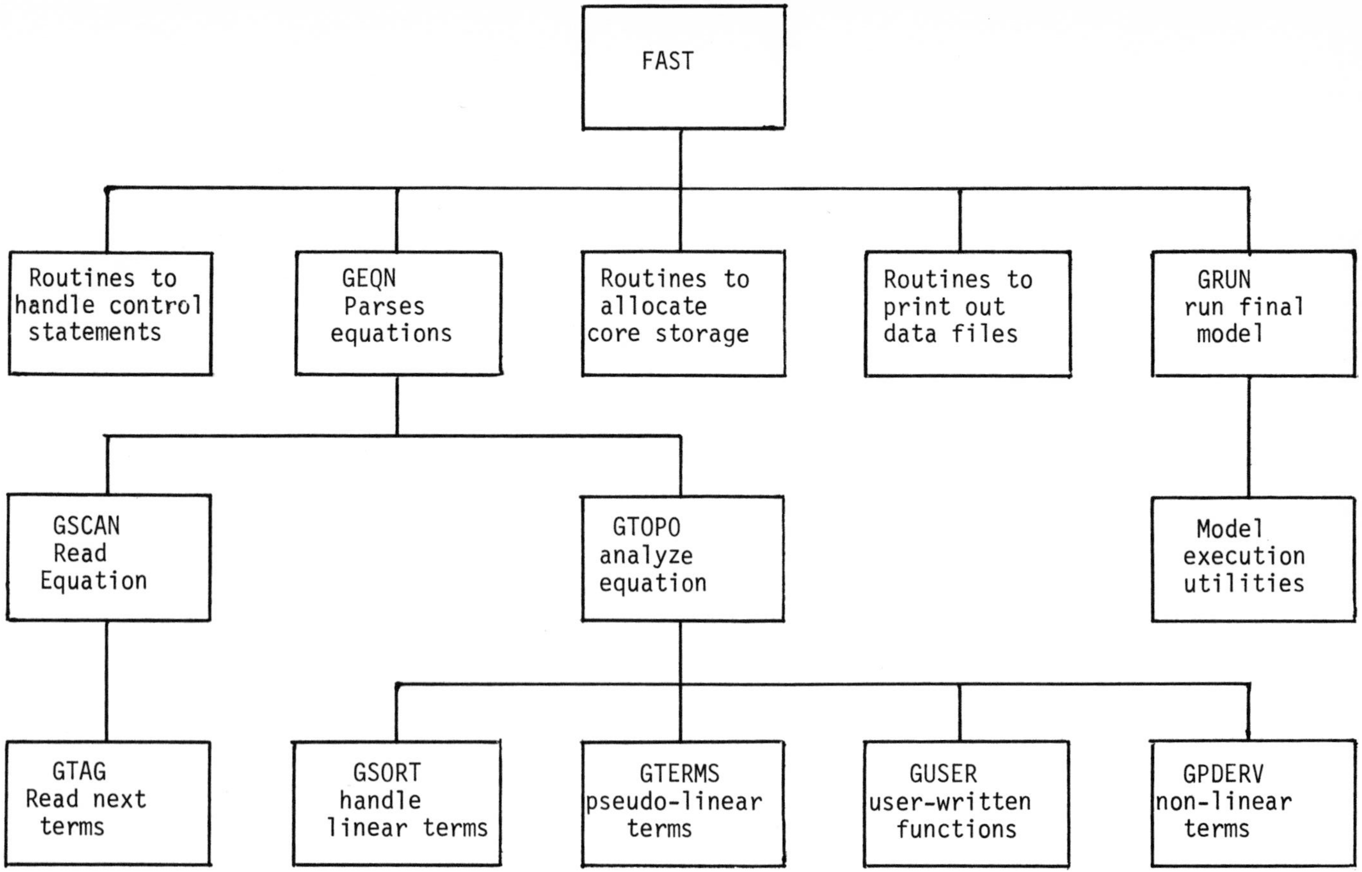
FAST
Routines to handle control statements
GEQN Parses equations
Routines to allocate core storage
Routines to print out data files
GRUN run final model
GSCAN Read Equation
GTOPO analyze equation
Model execution utilities
GTAG Read next terms
GSORT handle linear terms
GTERMS pseudo-linear terms
GUSER user-written functions
GPDERV non-linear terms

$$\frac{dX_2}{dX_4} = -2. * \text{FUN } (X_1, X_2)$$

$$\frac{dX_3}{dX_4} = 0.1 * \text{FUN } (X_1, X_2)$$

where FUN $(X_1, X_2) = \dfrac{X_1 * X_2}{(0.2 + X_2)}$ is a user written function put after the end card of the program. For this example the automatic step option is used. The perturbation size is proportional to DELTA = 10^{-3}.

The FAST program is as follows:

```
NUMBER -- STATEMENT

          ***
          *  EXAMPLE 1
          *
          *  A BIOLOGICAL REACTOR
          ***
   1         DX1 = FUN(X1,X2)-.4*X1*X3;
   2         DX2 = FUN(X1,X2)*(-2);
   3         DX3 = FUN(X1,X2)*(.1);
          ***
   4         INITIAL, X1=.1, X2=10;
   5         FINISH, X0.GE.2;
   6         PARAMETERS, DIRECT=RETAIN, STEP=0.1, OUTPUT=.2, ITER=20,
                         DELTA=1E-3, LIMIT=2000;
   7         PRINT, X0, X1, X2, X3;
   8         END;
                FUNCTION FUN(X1,X2)
                FUN=X1*X2/(0.2+X2)
                RETURN
                END
```

The results are:

```
        X0               X1               X2               X3

      0.0            1.00000E-01      1.00000E 01      0.0
WM03 -- WARNING MESSAGE....  MINIMUM STEP SIZE USED.
      2.00001E-01    1.21627E-01      9.95654E 00      2.16581E-03
      4.00001E-01    1.47914E-01      9.90377E 00      4.79954E-03
      6.00001E-01    1.79841E-01      9.83959E 00      8.00152E-03
      8.00001E-01    2.18599E-01      9.76163E 00      1.18938E-02
      1.00000E 00    2.65616E-01      9.66698E 00      1.66235E-02
      1.20000E 00    3.22600E-01      9.55205E 00      2.23683E-02
      1.40000E 00    3.91590E-01      9.41253E 00      2.93417E-02
      1.60000E 00    4.75011E-01      9.24334E 00      3.78007E-02
      1.80000E 00    5.75724E-01      9.03827E 00      4.80530E-02
      2.00000E 00    6.97085E-01      8.79001E 00      6.04661E-02

RUN TERMINATED -- FINISH CONDITION SATISFIED
```

Example 2: A Non-Ideal, Non-Linear Reactor System

This system consists of two CSTR's in series with a plug flow reactor. The reaction is represented by the irreversible non-linear kinetics:

$$3A \rightarrow B$$

$$r_A = kC_A^{\,3}$$

FLOW DIAGRAM

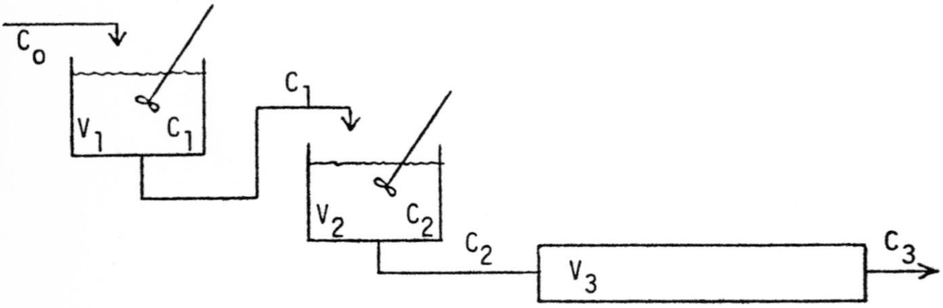

Figure 2

The defining equations for ideal reactors are:

$$\text{CSTR \#1} \quad \frac{dC_1}{dt} + kC_1^3 = F/V_1\,(C_0 - C_1)$$

$$\text{CSTR \#2} \quad \frac{dC_2}{dt} + kC_2^3 = F/V_2\,(C_1 - C_2)$$

$$\text{Plug flow reactor:} \quad \frac{\partial C_3}{\partial t} + F/A\,\frac{\partial C_3}{\partial Z} = kC_3^3$$

For this example the plug flow reactor was broken down into 20 cells.

The system is forced by a sine function.

The problem is coded in FAST as:

```
NUMBER -- STATEMENT

          ***
          *  EXAMPLE 2
          *
          *  A NON-IDEAL, NON-LINEAR REACTOR SYSTEM
          ***
     1       CONSTANT, F=50, CO=.8, K=.3, N=20;
     2       ARRAY, V=4,1,20;
     3       COMPUTE, DV=V(3)/N;
     4       FORCE, S=SINE(2,1.5,2);
     5       X1 = CO+S(XO)*.2;
     6       DX2 = F/V(1)*(X1-X2)-K*(X2**3);
     7       DX3 = F/V(2)*(X2-X3)-K*(X3**3);
          ***
     8       DX 4 = F/DV*(X 3-X 4)-K*(X 4**3);
     9       DX 5 = F/DV*(X 4-X 5)-K*(X 5**3);
    10       DX 6 = F/DV*(X 5-X 6)-K*(X 6**3);
    11       DX 7 = F/DV*(X 6-X 7)-K*(X 7**3);
    12       DX 8 = F/DV*(X 7-X 8)-K*(X 8**3);
    13       DX 9 = F/DV*(X 8-X 9)-K*(X 9**3);
    14       DX10 = F/DV*(X 9-X10)-K*(X10**3);
    15       DX11 = F/DV*(X10-X11)-K*(X11**3);
    16       DX12 = F/DV*(X11-X12)-K*(X12**3);
    17       DX13 = F/DV*(X12-X13)-K*(X13**3);
    18       DX14 = F/DV*(X13-X14)-K*(X14**3);
    19       DX15 = F/DV*(X14-X15)-K*(X15**3);
    20       DX16 = F/DV*(X15-X16)-K*(X16**3);
    21       DX17 = F/DV*(X16-X17)-K*(X17**3);
    22       DX18 = F/DV*(X17-X18)-K*(X18**3);
    23       DX19 = F/DV*(X18-X19)-K*(X19**3);
    24       DX20 = F/DV*(X19-X20)-K*(X20**3);
    25       DX21 = F/DV*(X20-X21)-K*(X21**3);
    26       DX22 = F/DV*(X21-X22)-K*(X22**3);
    27       DX23 = F/DV*(X22-X23)-K*(X23**3);
          ***
    28       INITIAL, XO=-5, X1=.8;
    29       FINISH, XO.GE.0;
    30       PARAMETERS, STEP=.1, OUTPUT=(.5,.1);
    31       MODIFY, CONT;
    32       FINISH, XO.GE.10;
    33       CALCOMP, X1, X2, X3, X23/XO = 'TIME';
    34       PRINT, XO, X1, X2, X3, X23;
    35       END;
```

The results are:

X0	X1	X2	X3	X23
8.37445E-05	8.00000E-01	7.88246E-01	7.85339E-01	7.33204E-01
5.00084E-01	8.00000E-01	7.88245E-01	7.85339E-01	7.33204E-01
1.00008E 00	8.00000E-01	7.88245E-01	7.85339E-01	7.33204E-01
1.50008E 00	8.00000E-01	7.88246E-01	7.85340E-01	7.33204E-01
2.00008E 00	9.81849E-01	8.13802E-01	7.96505E-01	7.33203E-01
2.50008E 00	8.76308E-01	8.79456E-01	8.80069E-01	8.05542E-01
3.00008E 00	7.29820E-01	7.42229E-01	7.45275E-01	7.87706E-01
3.50008E 00	6.20990E-01	6.27784E-01	6.29579E-01	6.77285E-01
4.00008E 00	6.08222E-01	5.99249E-01	5.97239E-01	5.90141E-01
4.50008E 00	6.98366E-01	6.72773E-01	6.66546E-01	5.81105E-01
5.00008E 00	8.43048E-01	8.07675E-01	7.98983E-01	6.55124E-01
5.50008E 00	9.64630E-01	9.30364E-01	9.21884E-01	7.67101E-01
6.00008E 00	9.97867E-01	9.76094E-01	9.70536E-01	8.53077E-01
6.50008E 00	9.24925E-01	9.21774E-01	9.20855E-01	8.72826E-01
7.00008E 00	7.84945E-01	7.95574E-01	7.98127E-01	8.22642E-01
7.50008E 00	6.53043E-01	6.63689E-01	6.66369E-01	7.19928E-01
8.00008E 00	6.00002E-01	5.97734E-01	5.97312E-01	6.16378E-01
8.50008E 00	6.54284E-01	6.34554E-01	6.29784E-01	5.75055E-01
9.00008E 00	7.86760E-01	7.53948E-01	7.45903E-01	6.19166E-01
9.50008E 00	9.26341E-01	8.90232E-01	8.81333E-01	7.24674E-01
1.00001E 01	9.98125E-01	9.70419E-01	9.63511E-01	8.26874E-01

RUN TERMINATED -- FINISH CONDITION SATISFIED

These results are plotted in Figure 3.

In addition to these, other examples are available (11,12, 13).

CONCLUSIONS

The FAST translator was designed to allow a user to obtain a solution to a problem directly from the modeling equations, thus freeing him from such tasks as dimensioning arrays, setting up loops, and checking convergence or errors, as would be required in FORTRAN. The model statements are the algebraic and first order ordinary differential equations taken directly from the problem definition, and any control options take the form of easy to understand and program control statements. The user need not know how to program even a simple problem in FORTRAN, as FAST writes all the programs required by IMP directly.

Work is continuing on the translator in an attempt to reduce even more the effort required to use it. User defined names will be allowed for state variables as well as constants, and a new, more efficient derivative analyzer using advanced data structures is under development. Then, a routine will be provided to accept higher order differential equations and produce a set of first order ones. Features similar to FORTRAN's CALL statement, to allow use of previously written routines, and DO statement, to ease the differencing of partial differential equations into a set of ordinary ones, are being considered. Other proposed features will be examined in the hopes of developing the most powerful tool possible, without requiring a substantial amount of programming knowledge on the part of the user.

Figure 3

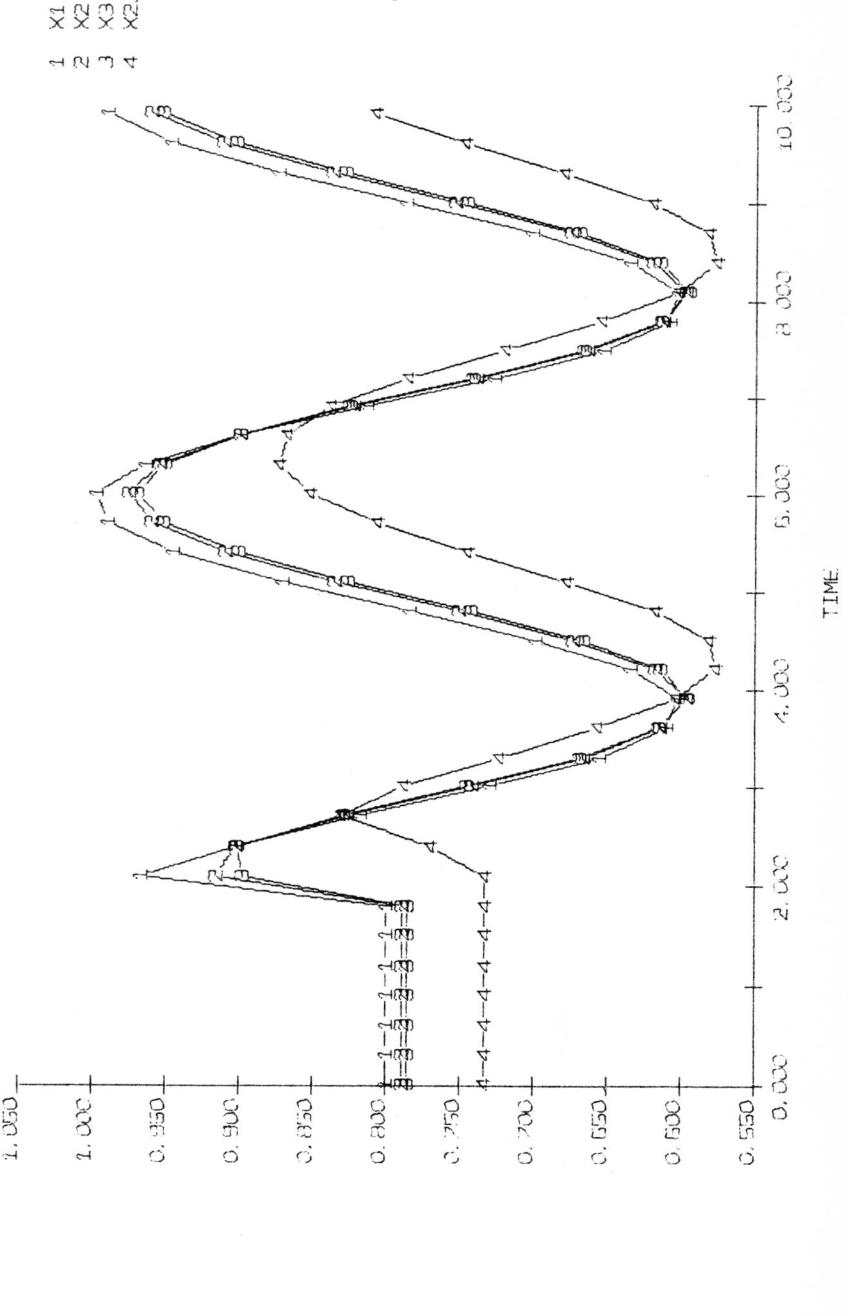
1 X1
2 X2
3 X3
4 X23
TIME
0.000
2.000
4.000
6.000
8.000
10.000
0.550
0.600
0.650
0.700
0.750
0.800
0.850
0.900
0.950
1.000
1.050

Nomenclature

A = reactor cross sectional area

$\bar{\bar{A}}$ = coefficient matrix

$\bar{\bar{A}}^S$ = Jacobian evaluated around expansion vector $\bar{Y}^S$ or expansion point X'

$\bar{B}$ = input vector

$\bar{B}^S$ = augmented input vector

C = composition

D = diffusivity coefficient

D = (as a superscript) diagonal

F = feed rate

h = integration step

k = reaction coefficient

V = reactor volume

$\bar{X}$ = state vector

X_n = state variable

Z_i = diagonal transition matrix

ΔZ = length increment

BIBLIOGRAPHY

1. Brandon, Daniel, Ph.D. Thesis, University of Connecticut, '73
2. Brandon, Daniel, "IMP Manual", University of Connecticut or Control Data Corporation
3. Stutzman, L. F., "An Efficient Procedure for the Simulation and Optimization of Chemical Systems Involving Differential Equations", Emploi des Calculateurs Electronique in Genie Chimique, European Federation of Chemical Engineering, Paris, April 24-28, 1973
4. Brandon, Daniel, "A New Single-Step Implicit Integration Algorithm with A-Stability and Improved Accuracy", Simulation 17-29, July 1974
5. Seinfeld, J., et al., Review of Numerical Integration Techniques for Stiff Ordinary Differential Equations, Ind. Eng. Chem. Fundamentals 9, #2, 1970
6. Factor, R. M., A Class of Implicit Methods for the Solution of Stiff Initial Value Problems, Progress Report #28-11, Systems Research Center, Case Western Reserve Univ., 1970
7. Calahan, D. S., et al., Stability Analysis of Numerical Integration, Proceedings of 10th Midwest Symposium on Circuit Theory, Lafayette, Indiana, May 1967
8. Leinger, W., et al., Efficient Numerical Integration of Stiff Systems of Ordinary Differential Equations, IBM Research Report RL-1970, 1967
9. Brandon, Daniel, "The Implementation and Use of Sparse Matrix Techniques in General Simulation Programs", The Computer Journal 17, #2, 1974
10. Stutzman, L. F., "Solution of Large Stiff Systems", Italian Conference on Use of Computers for Solving Process Problems, Pisa, Italy, May 6-7, 1974
11. Morgan, Richard, "FAST, The IMP Continuous System Translator", M.S. Project, University of Connecticut, 1974
12. Deschard, Frederic, "FAST, A Continuous System Translator for Differential Equations", M.S. Thesis, University of Connecticut, 1974
13. Stutzman, L. F., Personal Files

ACKNOWLEDGEMENTS

The authors express their appreciation to Paul Babcock, Post-doctorate in Chemical Engineering, for his contribution to the analytical differentiation routine, to him and many undergraduate and graduate students who used FAST in the solution of problems for their comments, suggestions, and criticisms, and to the Computer Center at the University of Connecticut for use of their facilities. The authors further wish to express thanks to Dr. Daniel Brandon for his support and critical review, and to Mr. William Norris, President of the Control Data Corporation; and to the Control Data Corporation for their support of the project through financial grants, a CDC 200 UT terminal, and access to CDC's CYBERNET Computer Systems.

Applications of EPISODE: An Experimental Package for the Integration of Systems of Ordinary Differential Equations

A. C. Hindmarsh and G. D. Byrne

1. Introduction

A standard class of problems, for which considerable literature and software exists, is that of initial value problems for first-order systems of ordinary differential equations. Using t as the independent variable (which usually, but not necessarily, denotes time), we can write the problem as

$$\dot{y} = f(y,t) \quad , \qquad y(t_o) = y_o \quad . \tag{1}$$

Here y, $\dot{y} = dy/dt$, and f are vectors of length N ($N \geqslant 1$), the initial value vector y_o and the function f(y,t) are given, and y(t) is to be computed on some interval (finite or infinite) one of whose endpoints is t_0. The problem may be stiff, i.e., have inherent time constants that are widely spread out, or it may have discontinuities or any of a number of other properties that can make the solution of (1) a highly nontrivial task.

The EPISODE program is a package of FORTRAN subroutines aimed at the automatic solution of problems of the form (1), with a minimum of effort required in the face of potential difficulties in the problem. The program implements both a generalized Adams method, well suited for nonstiff problems, and a generalized backward differentiation formula (BDF), well suited for stiff problems. Both methods are of implicit linear multistep type. In solving stiff problems, the package makes heavy use of the $N \times N$ Jacobian matrix,

---

Work performed under the auspices of the U.S. Energy Research and Development Administration. Contract #7405-ENG-48.

$$J = \frac{\partial f}{\partial y} = \left(\frac{\partial f_i}{\partial y_j}\right)_{i,j=1}^{N} \tag{2}$$

(the f_i and y_j are the vector components of f and y, respectively). But it is capable of approximating J internally if it is not feasible for the user to supply it directly. The methods used are discussed more fully in [1] and [4].

From the user's point of view, EPISODE very much resembles a package called GEAR [5], which is a heavily modified form of C.W. Gear's well-known code, DIFSUB [2, pp. 158-166]. There is a fundamental difference, however. The GEAR package is based on fixed-step formulas (Adams and BDF), and achieves changes in step size (when required) by interpolating to generate the multipoint data needed at the new spacing. In contrast, EPISODE is based on formulas that are truly variable-step, and step size changes can occur as frequently as every step, with no interpolation involved. (Interpolation is done to yield output at prescribed values of t, but this has no effect on the continuing integration process.) Like GEAR, EPISODE varies its order in a dynamic way, as well as its step size, in an effort to complete the integration with a minimum number of steps.

The difference in behavior between GEAR and EPISODE was shown dramatically at LLL in a class of chemistry problems related to atmospheric simulation. Here the function f involved kinetic rate constants which varied diurnally, i.e., with the rising and setting of the sun in the simulation code. This variation was often so sharp as to appear to be nearly a square wave. On these problems, the GEAR package generally behaved unstably, with step sizes oscillating wildly, and frequently terminating after repeated failures of the internal error tests. The EPISODE package, howver, has performed well on problems of this type, with up to 30 species, and time spans of months. A simple example of such a problem is given below (Example 2).

A complete discussion of the use of EPISODE is given in [4] and need not be repeated here. However, a few basic parameter definitions are needed here, in order to present the examples in the section which follows. Beyond the specification of the problem itself, represented by (1) and perhaps (2), the most important input parameter to EPISODE is the method flag, MF. This has eight values — 10, 11, 12, 13, 20, 21, 22, and 23. The first digit of MF, called METH, indicates the basic method to be used:

METH = 1 means implicit Adams;

METH = 2 means BDF.

The second digit, called MITER, indicates the method of iterative solution of the implicit equations arising from the chosen formula:

MITER = 0 means functional (or fixed-point) iteration (no Jacobian matrix used);

MITER = 1 means a chord method (or generalized Newton method, or semi-stationary Newton iteration) with Jacobian given by a subroutine supplied by the user;

MITER = 2 means a chord method with Jacobian generated internally by finite differences;

MITER = 3 means a chord method with a diagonal approximation to the Jacobian, generated internally (at less cost in storage and computation, but with reduced effectiveness).

The user must pick a value for MF, either based on some advance knowledge of the stiffness of the problem, or by experimentation with various values. The logical choice for non-stiff problems is MF = 10, and for stiff problems it is either 21 or 22. However, when J is quite diagonally dominant, MF = 13 or 23 may be best, and for mildly stiff problems, MF = 11 or 12 may be best.

Two other input parameters, EPS and IERROR, determine the nature of the error control performed within EPISODE. The constant EPS is the user's bound on the estimated local errors, normally measured on an error-per-step basis (but with other possibilities easily available). It must be remembered that the (global) errors in the final solution arise from an accumulation of local errors, and so EPS may need to be somewhat smaller than the user's tolerance limit on the final errors. Thus, for example, it may be necessary to use EPS = 10^{-6} in order to limit global errors to 10^{-4}. The manner in which errors in the individual components $y_i(t)$ are measured is determined by IERROR:

IERROR = 1 means control of absolute errors;

IERROR = 2 means control of purely relative errors;

IERROR = 3 means control of errors in a semi-relative sense — relative to the largest value of $|y_i(t)|$ seen so far, or to 1 if $y_i(t_o) = 0$ and $|y_i|$ has remained below 1.

Another required input parameter is the initial step size to be attempted, HØ. The value of this is not crucial, as the code will reduce it if necessary for the error tests, and/or later increase it as appropriate for the sake of efficiency.

The EPISODE package is used by making calls to a driver subroutine, DRIVE, which in turn calls other routines in the package to solve the problem. The function f is communicated by way of a subroutine, DIFFUN, which the user must write. If MITER = 1, a subroutine for the Jacobian, PEDERV, must also be written. Calls to DRIVE are made repeatedly, once for each of the user's output points. A value of t at which output is desired is put in the argument TOUT to DRIVE, and when (or if) TOUT is reached, control returns to the calling program with the value of y at t = TOUT. Another argument to DRIVE, called INDEX, is used to convey whether or not the call is the first one for the problem (and thus whether to initialize various variables). It is also used as an output argument, to convey the success or failure of the package in performing the requested task.

The following section discusses in some detail three example problems which can be solved by EPISODE. These attempt to show the capabilities of the package in a variety of situations, with the use of a variety of the above input options. Following that is a short description of the package itself, including its availability.

2. Examples

In order to illustrate how the EPISODE package can be used to solve non-trivial initial value problems, we give here a few examples, chosen from the areas of chemical kinetics and atmospheric modeling. For each example problem, the appropriate FORTRAN coding for its solution, with EPISODE, is given, followed by the output generated by that coding. The computer runs for these examples were made on a CDC 7600 machine at LLL, a machine with about 14 decimal figures of accuracy.

Example 1. A kinetics problem. The following chemical kinetics problem, given by Robertson, is frequently used as an illustrative example [7, pp. 268-269]. It involves the following three nonlinear rate equations:

$$\dot{y}_1 = -.04\ y_1 + 10^4\ y_2 y_3$$
$$\dot{y}_2 = .04\ y_1 - 10^4\ y_2 y_3 - 3\cdot 10^7\ y_2^2$$
$$\dot{y}_3 = 3\cdot 10^7\ y_2^2\ .$$

(Note that we do not rescale this problem, to eliminate the wide spread of coefficient values and solution values, as is often done. We feel that such a procedure is usually an unnecessary inconvenience when a quality solver is being used, and in practice is not generally feasible until the solution is known.) The initial values at t = 0 are

$$y_1(0) = 1\ , \qquad y_2(0) = y_3(0) = 0\ .$$

Since $\Sigma \dot{y}_i = 0$, the solution must satisfy $\Sigma y_i = 1$ identically. This identity can be used as an error check, but it is not a good one, because, for most solution algorithms, $\Sigma y_i - 1$ will remain at the level of the machine unit roundoff (10^{-14} here) while the inaccuracies in the y_i themselves are much larger (on the order of EPS).

Suppose we wish to solve this problem with the BDF method (METH = 2), and use the chord iteration method with user-supplied Jacobian (MITER = 1), i.e., MF = 21. Suppose also we choose a local error bound of EPS = 10^{-6}, and control absolute error (IERROR = 1). We choose an initial step size of HØ = 10^{-8}. The use of MITER = 1 requires that the Jacobian $J = \partial f/\partial y$ be calculated and programmed. This is given by

$$J = \begin{pmatrix} -.04 & 10^4 y_3 & 10^4 y_2 \\ .04 & -10^4 y_3 - 6\cdot10^7 y_2 & -10^4 y_2 \\ 0 & 6\cdot10^7 y_2 & 0 \end{pmatrix}$$

From physical reasoning, we would know in advance that the solution of this problem approaches equilibrium as $t \to \infty$, but we would not know in advance how far to integrate to observe this equilibrium numerically. The final value of t given in [7] is 40. So consider taking output at $t = 4\cdot10^k$, $k = -1, 0, 1, 2, \ldots$. These will be the values of the argument TOUT. If we were running the problem interactively, we might look at each output line before deciding whether or not to terminate. However, for the runs given here, we simply take $k = -1, 0, 1, \ldots, 10$, but terminate if $|\Sigma y_i - 1|$ ever exceeds EPS.

The following coding, together with the EPISODE package, can be used to solve this problem with the options described above. The calling program prints a line of output at each output point. This line includes t, the y_i, and $\Sigma y_i - 1$. By accessing one of the Common blocks of the package, we may also obtain, at each output point, the following statistics:

NSTEP = the number of steps taken so far;
NFE = the number of f evaluations so far;
NJE = the number of J evaluations so far;
NQUSED = the order (NQ) last used;
HUSED = the step size (H) last used (also available as the returned value of HØ).

Notice that the updated solution vector y(t) and the corresponding value of t are returned in YØ and TØ, respectively, which were used to supply the initial conditions. Also, an error halt occurs if a nonzero value of INDEX is returned. Notice that the coding for DIFFUN makes use of the conservation law $\sum \dot{y}_i = 0$, and similar relations among the elements of the Jacobian are utilized in the coding of PEDERV. Also, notice that the array, PD, in which PEDERV stores the Jacobian, is dimensioned with a column length of NØ, not N. Normally NØ will be equal to N, but the case NØ > N is easily allowed for, possibly with a minor change to the driver routine, DRIVE. The reason for this is that in chemical kinetics calculations, it is often convenient to store constant concentrations on the end of the vector of dynamic ones, or to reduce the number of dynamic species during the problem. The latter capability is provided for in EPISODE with no change to the package.

```
C  PROGRAM FOR EXAMPLE PROBLEM 1
      DIMENSION Y0(3)
      COMMON /EPCOM9/ HUSED,NQUSED,NSTEP,NFE,NJE
      N = 3
      T0 = 0.
      H0 = 1.E-8
      Y0(1) = 1.
      Y0(2) = 0.
      Y0(3) = 0.
      EPS = 1.E-6
      IERROR = 1
      MF = 21
      WRITE(3,20)MF,EPS
 20   FORMAT(//49H EXAMPLE PROBLEM 1.. ROBERTSON 3-SPECIES CHEMICAL,
     1      17H KINETICS PROBLEM//5H MF =,I3,8H   EPS =,E10.1///
     2       4H   T,6X,5HNSTEP,3X,3HNFE,3X,3HNJE,2X,2HNQ,4X,1HH,10X,
     3       2HY1,10X,2HY2,10X,2HY3,5X,9H SUM(Y)-1//)
      NOUT = 12
      TOUT = .4
      INDEX = 1
      DO 60 IOUT = 1,NOUT
        CALL DRIVE(N,T0,H0,Y0,TOUT,EPS,IERROR,MF,INDEX)
        SY = Y0(1) + Y0(2) + Y0(3) - 1.
        WRITE(3,40)T0,NSTEP,NFE,NJE,NQUSED,H0,Y0(1),Y0(2),Y0(3),SY
 40     FORMAT(E9.1,3I6,I4,E10.2,3E12.4,E9.1)
        IF (ABS(SY) .GT. EPS) GO TO 80
        IF (INDEX .NE. 0) GO TO 90
 60     TOUT = TOUT*10.
 80   CALL EXIT
 90   WRITE(3,100)INDEX
 100  FORMAT(///21H ERROR HALT   INDEX =,I3)
      CALL EXIT
      END


      SUBROUTINE DIFFUN(N,T,Y,YDOT)
      DIMENSION Y(1),YDOT(1)
      YDOT(1) = -.04*Y(1) + 1.E4*Y(2)*Y(3)
      YDOT(3) = 3.E7*Y(2)*Y(2)
      YDOT(2) = -YDOT(1) - YDOT(3)
      RETURN
      END


      SUBROUTINE PEDERV(N,T,Y,PD,N0)
      DIMENSION Y(1),PD(N0,1)
      PD(1,1) = -.04
      PD(1,2) = 1.E4*Y(3)
      PD(1,3) = 1.E4*Y(2)
      PD(2,1) = .04
      PD(2,3) = -PD(1,3)
      PD(3,1) = 0.
      PD(3,2) = 6.E7*Y(2)
      PD(3,3) = 0.
      PD(2,2) = -PD(1,2) - PD(3,2)
      RETURN
      END
```

The output of the above program is as follows:

EXAMPLE PROBLEM 1.. ROBERTSON 3-SPECIES CHEMICAL KINETICS PROBLEM

MF = 21 EPS = 1.0E-06

T	NSTEP	NFE	NJE	NQ	H	Y1	Y2	Y3	SUM(Y)-1
4.0E-01	27	38	19	3	1.63E-01	9.8517E-01	3.3864E-05	1.4794E-02	-1.1E-13
4.0E+00	40	60	23	5	5.61E-01	9.0552E-01	2.2405E-05	9.4462E-02	-1.7E-13
4.0E+01	89	152	30	4	2.31E+00	7.1582E-01	9.1851E-06	2.8417E-01	-4.0E-13
4.0E+02	130	233	38	4	2.01E+01	4.5051E-01	3.2228E-06	5.4949E-01	-6.5E-13
4.0E+03	180	327	50	5	2.37E+02	1.8320E-01	8.9423E-07	8.1680E-01	-9.0E-13
4.0E+04	225	410	66	5	3.26E+03	3.8986E-02	1.6219E-07	9.6101E-01	-1.1E-12
4.0E+05	256	474	71	4	2.92E+04	4.9409E-03	1.9839E-08	9.9506E-01	-1.2E-12
4.0E+06	295	559	84	4	1.85E+05	5 0319E-04	2.0117E-09	9.9950E-01	-1.3E-12
4.0E+07	319	595	91	3	4.31E+06	5.2329E-05	2.0934E-10	9.9995E-01	-1.4E-12
4.0E+08	332	612	99	4	7.21E+07	4.7709E-06	1.9084E-11	1.0000E+00	-1.5E-12
4.0E+09	341	621	105	3	1.23E+09	5.4561E-07	2.1824E-12	1.0000E+00	-1.3E-12
4.0E+10	352	652	115	3	2.86E+09	-1.1761E-06	-4.7043E-12	1.0000E+00	-7.2E-12

We see that the equilibrium values are

$$y_1 = y_2 = 0 \quad , \qquad y_3 = 1 \quad ,$$

and that the approach to equilibrium is quite slow. Notice, however, how the time step, H, rises steadily with time, T. We also observe that the code generated negative and thus physically incorrect answers during the last decade. A closer study shows that this reflects an instability, or a high sensitivity of the problem to numerical errors at late t, and will, if the integration is continued, lead to answers diverging to $\pm\infty$. We can see this by noting that J, evaluated at the equilibrium values, has eigenvalues of 0, 0, and $-10^{-4} - .04$. Small errors in y can perturb at least one of the zero eigenvalues to positive values, thereby leading to unbounded growth in the numerical solution, according to the usual linear stability theory.

The accuracy of the above solution (prior to the instability) can be verified in the usual way — by rerunning the problem with a smaller value of EPS. With EPS = 10^{-9} and nothing else changed, the output is as follows:

EXAMPLE PROBLEM 1.. ROBERTSON 3-SPECIES CHEMICAL KINETICS PROBLEM

MF = 21 EPS = 1.0E-09

T	NSTEP	NFE	NJE	NQ	H	Y1	Y2	Y3	SUM(Y)-1
4.0E-01	78	119	28	4	3.42E-02	9.8517E-01	3.3864E-05	1.4794E-U2	-2.8E-13
4.0E+00	115	190	33	5	1.43E-01	9.0552E-01	2.2405E-05	9.4459E-02	-4.4E-13
4.0E+01	199	358	51	5	1.27E+00	7.1583E-01	9.1855E-06	2.8416E-01	-8.5E-13
4.0E+02	314	586	84	5	1.01E+01	4.5052E-01	3.2229E-06	5.4948E-01	-1.5E-12
4.0E+03	422	777	102	5	8.46E+01	1.8320E-01	8.9424E-07	8.1680E-01	-2.1E-12
4.0E+04	520	942	112	5	1.10E+03	3.8983E-02	1.6218E-07	9.6102E-01	-2.5E-12
4.0E+05	607	1102	120	5	1.14E+04	4.9383E-03	1.9850E-08	9.9506E-01	-2.8E-12
4.0E+06	675	1227	129	5	2.20E+05	5.1681E-04	2.0683E-09	9.9948E-01	-3.0E-12
4.0E+07	728	1327	143	5	3.12E+06	5.2031E-05	2.0809E-10	9.9995E-01	-3.2E-12
4.0E+08	778	1422	155	3	1.94E+07	5.2096E-06	2.0839E-11	9.9999E-01	-3.4E-12
4.0E+09	808	1473	163	4	4.39E+08	5.1925E-07	2.1150E-12	1.0000E+00	-3.5E-12
4.0E+10	828	1508	170	3	7.22E+09	5.2003E-08	2.0721E-13	1.0000E+00	-3.6E-12

Note that here the instability did not arise in the t range used, because the numerical errors were kept smaller. The absolute differences between corresponding solution values in the two runs are all less than about 10^{-5}, indicating that, at least in the first run, the accuracy is consistent with the value of EPS used.

For a similar but more complex problem, a user might choose MF = 22 (Jacobian computed numerically internally) instead of 21. With this choice, and EPS = 10^{-6}, the output for this example is as follows:

EXAMPLE PROBLEM 1.. ROBERTSON 3-SPECIES CHEMICAL KINETICS PROBLEM

MF = 22 EPS = 1.0E-06

T	NSTEP	NFE	NJE	NQ	H	Y1	Y2	Y3	SUM(Y)-1
4.0E-01	27	95	19	3	1.63E-01	9.8517E-01	3.3864E-05	1.4794E-02	-9.9E-14
4.0E+00	40	129	23	5	5.61E-01	9.0552E-01	2.2405E-05	9.4462E-02	-1.6E-13
4.0E+01	77	225	29	4	2.50E+00	7.1582E-01	9.1849E-06	2.8417E-01	-3.3E-13
4.0E+02	115	311	34	4	1.83E+01	4.5052E-01	3.2229E-06	5.4948E-01	-5.3E-13
4.0E+03	148	408	45	5	2.25E+02	1.8321E-01	8.9426E-07	8.1679E-01	-7.2E-13
4.0E+04	191	542	58	5	2.06E+03	3.8987E-02	1.6224E-07	9.6101E-01	-9.1E-13
4.0E+05	231	633	65	4	3.04E+04	4.9394E-03	1.9854E-08	9.9506E-01	-1.1E-12
4.0E+06	265	722	74	3	3.96E+05	5.1164E-04	2.0476E-09	9.9949E-01	-1.2E-12
4.0E+07	327	982	117	1	4.25E+06	3.2146E-05	1.2859E-10	9.9997E-01	-1.5E-12
4.0E+08	775	2427	270	5	1.10E+07	-1.3361E+05	-4.0000E-06	1.3361E+05	9.3E-10
4.0E+09	861	2747	312	4	5.82E+07	-1.8616E+06	-4.0000E-06	1.8616E+06	2.2E-08
4.0E+10	923	2954	330	5	1.20E+09	-1.9142E+07	-4.0000E-06	1.9142E+07	0.

Note that here the instability discussed above arose much earlier, and shows more clearly the divergence to infinity. This is presumably because of inaccuracies in the Jacobian matrix. The values of NFE in the above output are considerably higher than at corresponding times in the run with MF = 21, because they include the cost of J evaluations. This cost is 3*NJE f evaluations, since each numerical evaluation of J involves N evaluations of f. If this cost is subtracted out, the result is the number of f evaluations used for corrector iterations only, and is about the same for MF = 22 as for 21. For reasons of both efficiency and accuracy, it is generally better to generate J analytically (MITER = 1) than numerically, however.

The analysis of the Jacobian mentioned above shows that this problem is stiff, because the smallest time constant is about 10^{-4} while the time span of the problem is many orders of magnitude larger. If the stiffness had not been known in this way, or on physical grounds, it could have been deduced simply by trying the nonstiff option, MF = 10. With this MF, EPS = 10^{-6}, and an inserted termination test on NFE, stopping the run when NFE exceeds 5000, the output is as follows:

EXAMPLE PROBLEM 1.. ROBERTSON 3-SPECIES CHEMICAL KINETICS PROBLEM

MF = 10 EPS = 1.0E-06

T	NSTEP	NFE	NJE	NQ	H	Y1	Y2	Y3	SUM(Y)-1
4.0E-01	1130	1928	0	1	2.61E-04	9.8517E-01	3.3837E-05	1.4793E-02	-5.0E-12
4.0E+00	11285	19333	0	1	5.67E-04	9.0552E-01	2.2432E-05	9.4457E-02	-5.3E-11

The excessive cost clearly indicates the stiffness here. In cases where the presence or absence of stiffness cannot be easily determined any other way, such numerical experimentation is an easy and direct way to determine it, provided care is taken not to overspend one's time using a nonstiff method on a stiff problem.

Example 2. A diurnal kinetics problem. A simple problem arising in atmospheric simulation [3] is the Chapman mechanism for ozone kinetics. This involves the oxygen singlet (O), ozone (O_3), and oxygen (O_2), and two of the reaction processes are photochemical, thus involving diurnal rate coefficients. We denote the concentrations of O, O_3 and O_2 by y_1, y_2, y_3, respectively, with the last one held at a constant value $y_3 = 3.7\cdot 10^{16}$ cm^{-3}. The initial values of the other two are (in cm^{-3})

$$y_1(0) = 10^6 \quad , \qquad y_2(0) = 10^{12} \quad .$$

The two rate equations are given by

$$\dot{y}_1 = -Q_1 - Q_2 + 2Q_3 + Q_4$$

$$\dot{y}_2 = Q_1 - Q_2 - Q_4$$

where

$$Q_1 = q_1\, y_1\, y_3 \quad , \qquad Q_2 = q_2\, y_1\, y_2 \quad ,$$

$$Q_3 = q_3(t)\, y_3 \quad , \qquad Q_4 = q_4(t)\, y_2 \quad .$$

The q_i are rate coefficients. Two of them are constant:

$$q_1 = 1.63\cdot 10^{-16} \quad , \qquad q_2 = 4.66\cdot 10^{-16} \quad .$$

The other two vary diurnally:

$$q_i(t) = \left\{ \begin{array}{ll} \exp(-c_i/\sin wt) \quad , & \sin wt > 0 \\ 0 \quad , & \sin wt \le 0 \end{array} \right\} \quad (i = 3,\ 4) \quad ,$$

$$c_3 = 22.62 \quad , \qquad c_4 = 7.601 \quad , \qquad w = \pi/43200 \quad ,$$

where t is in seconds. These diurnal functions are very nearly like square waves, rising sharply at dawn ($t = 0$), remaining near their peak (noontime) values in the daytime ($\sin wt > 0$), falling sharply at sunset ($t = 43200$) to a nighttime value of zero, and continuing periodically. It is these functions that make this problem, and others like it, a considerable challenge to solve numerically.

For this example, we run the problem for 10¼ days ($0 \leqslant t \leqslant 8.856\cdot 10^5$), and take output every 2 hours during the first 12 hours, then only at midnight and noon thereafter.

Again we could use either physical reasoning or experimentation to determine the stiffness of this problem. We could also look at the Jacobian matrix,

$$J = \begin{pmatrix} -q_1 y_3 - q_2 y_2 & -q_2 y_1 + q_4(t) \\ q_1 y_3 - q_2 y_2 & -q_2 y_1 - q_4(t) \end{pmatrix}$$

The initial value of J is approximately

$$J(0) \cong \begin{pmatrix} -6 & -5\cdot 10^{-10} \\ 6 & -5\cdot 10^{-10} \end{pmatrix}$$

and has eigenvalues of approximately -6 and -10^{-9}. The former corresponds to a time constant of 1/6 sec, much smaller than the time range of interest, and hence indicates stiffness. Therefore, we again choose the method option MF = 21.

This problem requires some thought as to the best type of error control. Since the values of y_i have large and widespread orders of magnitude, absolute error control is out of the question, and we consider relative error control (IERROR = 2). However, the 0 concentration, y_1, drops to extremely small values at night, as can be anticipated by the chemist, or discovered by numerical experimentation. When that happens, relative error control is overly demanding, and we need to modify it. The semi-relative control given by IERROR = 3 is not appropriate, because the largest value of $|y_1|$ seen will be the previous peak daytime value, which is too large to give meaningful control at night. (For example, with EPS = 10^{-6}, and a peak y_1 of, say, 10^7, local errors in y_1 would only be kept below about $10^{-6}\cdot 10^7 = 10$.) Suppose we choose to control errors in y_i relative to the quantity $\max(|y_i(t)|, 10^{-20})$. Then the resulting control will be on relative error when $|y_i|$ is above the floor value of 10^{-20}. When it is below, the control is relative to that floor value, i.e., it attempts to keep local errors less than EPS$\cdot 10^{-20}$. To achieve this type of error control, we need to access an array called YMAX, stored in the Common block EPCOM2 within the EPISODE package; the local error in y_i is controlled to be less than EPS*YMAX(i), approximately. We may, therefore, simply make the appropriate Common declaration and set the values of YMAX, to the expression given above, in an appropriate place. We may do this in Subroutine DIFFUN, for example, where it will take effect continually, as DIFFUN is called at least once per step. An alternative is to modify the coding in DRIVE where YMAX is updated, but we choose the first way for this example. We again choose EPS = 10^{-6} and HØ = 10^{-8}.

The following coding can be used to solve this problem, with the options discussed, using EPISODE. For each output time, the y vector and the same statistics as given in Example 1 are printed on a line.

```
C  PROGRAM FOR EXAMPLE PROBLEM 2
      DIMENSION Y0(2)
      COMMON /EPCOM9/ HUSED,NQUSED,NSTEP,NFE,NJE
      COMMON /PCOM/ Q1,Q2,C3,C4,FREQ,Y3
      N = 2
      T0 = 0.
      H0 = 1.E-8
      Y0(1) = 1.E6
      Y0(2) = 1.E12
      EPS = 1.E-6
      IERROR = 2
      MF = 21
      Q1 = 1.63E-16
      Q2 = 4.66E-16
      C3 = 22.62
      C4 = 7.601
      FREQ = 3.1415926535898/43200.
      Y3 = 3.7E16
      WRITE(3,20)MF,EPS
 20   FORMAT(//46H EXAMPLE PROBLEM 2.. DIURNAL CHEMISTRY PROBLEM//
     1      5H MF =,I3,8H    EPS =,E10.1///
     2       4H    T,8X,5HNSTEP,3X,3HNFE,3X,3HNJE,2X,2HNQ,4X,1HH,10X,
     3       2HY1,10X,2HY2//)
      NOUT = 26
      TOUT = 7200.
      INDEX = 1
      DO 60 IOUT = 1,NOUT
        CALL DRIVE(N,T0,H0,Y0,TOUT,EPS,IERROR,MF,INDEX)
        WRITE(3,40)T0,NSTEP,NFE,NJE,NQUSED,H0,Y0(1),Y0(2)
 40     FORMAT(E11.3,3I6,I4,E10.2,2E12.4)
        IF (INDEX .NE. 0) GO TO 90
        IF (IOUT .LT. 6) TOUT = TOUT + 7200.
        IF (IOUT .EQ. 6) TOUT = TOUT + 21600.
        IF (IOUT .GT. 6) TOUT = TOUT + 43200.
 60     CONTINUE
      CALL EXIT
 90   WRITE(3,100)INDEX
 100  FORMAT(///21H ERROR HALT    INDEX =,I3)
      CALL EXIT
      END

      SUBROUTINE DIFFUN(N,T,Y,YDOT)
      DIMENSION Y(1),YDOT(1)
      COMMON /EPCOM2/ YMAX(20)
      COMMON /PCOM/ Q1,Q2,C3,C4,FREQ,Y3
      CALL DIURN(T,Q3,Q4)
      QQ1 = Q1*Y(1)*Y3
      QQ2 = Q2*Y(1)*Y(2)
      QQ3 = Q3*Y3
      QQ4 = Q4*Y(2)
      YDOT(1) = -QQ1 - QQ2 + 2.*QQ3 + QQ4
      YDOT(2) = QQ1 - QQ2 - QQ4
      YMAX(1) = AMAX1(ABS(Y(1)),1.E-20)
      YMAX(2) = AMAX1(ABS(Y(2)),1.E-20)
      RETURN
      END

      SUBROUTINE PEDERV(N,T,Y,PD,N0)
      DIMENSION Y(1), PD(N0,1)
      CCMMON /PCOM/ Q1,Q2,C3,C4,FREQ,Y3
      CALL DIURN(T,Q3,Q4)
      PD(1,1) = -Q1*Y3 - Q2*Y(2)
      PD(1,2) = -Q2*Y(1) + Q4
      PD(2,1) = Q1*Y3 - Q2*Y(2)
      PD(2,2) = -Q2*Y(1) - Q4
      RETURN
      END
```

```
      SUBROUTINE DIURN(T,Q3,Q4)
      COMMON /PCOM/ Q1,Q2,C3,C4,FREQ,Y3
      S = SIN(FREQ*T)
      IF (S .LE. 0.) GO TO 20
      Q3 = EXP(-C3/S)
      Q4 = EXP(-C4/S)
      RETURN
 20   Q3 = 0.
      Q4 = 0.
      RETURN
      END
```

Several further differences between this program and that of Example 1 should be noted. First, the logic for setting the values of TOUT is more complicated. Secondly, a Common block, PCOM, is used to communicate problem dependent quantities to the subroutines, as set in the main program. Finally, an auxiliary subroutine, DIURN, is used by both DIFFUN and PEDERV.

The output of this code is as follows:

```
EXAMPLE PROBLEM 2.. DIURNAL CHEMISTRY PROBLEM

MF = 21    EPS =    1.0E-06
```

T	NSTEP	NFE	NJE	NQ	H	Y1	Y2
7.200E+03	934	116S	89	5	1.73E+02	4.1428E+04	1.0000E+12
1.440E+04	966	1227	98	4	6.63E+02	2.5635E+07	1.0002E+12
2.160E+04	988	1272	105	5	4.85E+02	8.7935E+07	1.0386E+12
2.880E+04	1008	1315	110	4	1.39E+02	2.7607E+07	1.0772E+12
3.600E+04	1044	1389	118	5	1.22E+02	4.4663E+04	1.0774E+12
4.320E+04	1455	2077	155	2	4.71E+02	3.0927E-33	1.0774E+12
6.480E+04	1462	2084	162	1	8.05E+03	-1.2178E-57	1.0774E+12
1.080E+05	1963	2837	230	5	4.16E+02	9.4344E+07	1.1160E+12
1.512E+05	2491	3698	315	1	1.11E+04	3.7062E-66	1.1547E+12
1.944E+05	2981	4432	377	5	4.55E+02	1.0074E+08	1.1932E+12
2.376E+05	3501	5340	437	2	4.28E+03	7.2072E-29	1.2318E+12
2.808E+05	3985	6052	488	5	4.33E+02	1.0712E+08	1.2702E+12
3.240E+05	4554	7056	596	1	8.28E+03	1.2209E-62	1.3087E+12
3.672E+05	5042	7772	652	5	5.50E+02	1.1349E+08	1.3470E+12
4.104E+05	5531	8610	713	1	8.22E+03	8.8202E-69	1.3855E+12
4.536E+05	6016	9328	768	5	5.49E+02	1.1984E+08	1.4237E+12
4.968E+05	6526	10198	831	1	8.34E+03	3.9297E-45	1.4620E+12
5.400E+05	7020	10923	881	4	3.03E+02	1.2618E+08	1.5001E+12
5.832E+05	7529	11795	946	1	7.70E+03	6.7907E-69	1.5384E+12
6.264E+05	8016	12506	996	5	4.99E+02	1.3250E+08	1.5764E+12
6.696E+05	8489	13334	1050	1	7.87E+03	-1.2415E-57	1.6145E+12
7.128E+05	8995	14081	1109	5	6.05E+02	1.3880E+08	1.6524E+12
7.560E+05	9517	14988	1184	1	9.24E+03	4.0193E-59	1.6904E+12
7.992E+05	9985	15674	1231	5	5.40E+02	1.4508E+08	1.7282E+12
8.424E+05	10522	16607	1315	1	8.84E+03	-3.2477E-59	1.7661E+12
8.856E+05	11015	17341	1370	5	5.02E+02	1.5135E+08	1.8038E+12

The solution shows a slowly rising concentration of O_3 and a diurnally oscillating concentration of O, with slowly rising noontime values. Note that the nighttime values of y_1 are essentially zero, but that the computed values include some small negative numbers. These are unphysical, but should cause no concern, since they are well within the requested error tolerance.

The order of the method, NQ, also oscillates in this problem. Low orders (1 or 2) are chosen at night, where there is little activity, and high orders (4 or 5) are chosen during the day, where there is considerable activity. This represents the code's attempt to choose the optimal order at all times, and shows the value of the variable-order feature.

The time step size, H, is seen to vary from about 2 minutes to over 3 hours. In fact, considerably smaller values of H, not seen here, are needed in the neighborhood of sunrise and sunset. For other problems of this type, it is possible, depending on the nature of the diurnal coefficients, for H to become so small and the time T so large that H/T is below the level of the unit round-off, and temporarily T+H = T on the machine. If this happens in EPISODE, a warning message is printed, but the run is not halted, nor should it be.

As before, we can estimate the accuracy of the above results by running the problem again with a smaller EPS. This was done, and again indicates that the results with EPS = 10^{-6} are accurate to five significant figures, except, of course, that the small nighttime values have only an absolute accuracy of 10^{-26} or better.

Example 3. A diffusion-convection problem. In this problem, posed as a model test problem by Sincovec [1, pp. 90-92], a partial differential equation is treated by the method of lines [8]. The equation, in u(x,t), is

$$\partial u/\partial t = \partial^2 u/\partial x^2 - c\ \partial u/\partial x \quad , \qquad 0 \leq x \leq 1 \quad , \qquad t \geq 0 \quad ,$$

where c is a constant. The boundary conditions posed are

$$u(0,t) = 1 \qquad (t > 0) \quad ,$$

$$\frac{\partial u}{\partial x}(1,t) = 0 \qquad (t > 0) \quad ,$$

$$u(x,0) = 0 \qquad (0 < x < 1) \quad .$$

We discretize the space variable x by breaking the unit interval into N equal subintervals, and approximating u(k/N,t) by a function $u_k(t)$ for k = 0, 1, ..., N. The u_k are then made to satisfy the ordinary differential equations obtained by central differencing:

$$\dot{u}_k = \frac{u_{k-1} - 2u_k + u_{k+1}}{(1/N)^2} - c\,\frac{u_{k+1} - u_{k-1}}{(2/N)} \quad , \qquad k = 1, 2, \ldots, N.$$

We simulate the initial and boundary conditions by setting

$$u_k(0) = 0 \quad (k = 1, 2, \ldots, N) \quad ,$$

$$u_o = 1 \text{ (all t)} \quad , \qquad u_{N+1} = u_{N-1} \text{ (all t)} \quad .$$

Thus we have an initial value problem in $y = (u_1, \ldots, u_N)^T$. The values used for this example are N = 20 and c = 25.

Physically, this problem can represent a simple model of a fluid flow, heat flow, or other phenomenon, in which an initially discontinuous profile is propagated by diffusion and convection, the latter with a speed of c. We therefore consider a time span of 1/c = .04, and take output at intervals of .1/c = .004. In order to show the movement of the wavefront, we print the values of u_1, $u_{10} = u_{N/2}$ and $u_{20} = u_N$ at each output time. Since there is no reason to suspect stiffness in advance, we consider MF = 10. The spatial discretization used introduces some error here, and so it is pointless to request too much accuracy in the time integration. We therefore take EPS = 10^{-3}, and choose the semi-relative error control, IERROR = 3. We choose an initial step size of HØ = 10^{-3}.

In passing, we note that this problem has the simple linear form $\dot{y} = Ay$ with a constant matrix A. Simple modifications can easily be made to the EPISODE code to take advantage of this form. Moreover, other methods, of a quite different nature, can be brought to bear on this class of problems. However, in the present context, this example serves only to represent more general partial differential equation problems, usually nonlinear. Such code specializations or special methods are then inapplicable, and we choose to ignore them here.

The following coding will solve this problem with the above options. Notice that a dummy Subroutine PEDERV is needed, even though it is never called, to satisfy the loader.

```
C     PROGRAM FOR EXAMPLE PROBLEM 3
      DIMENSION Y0(20)
      COMMON /EPCOM9/ HUSED,NQUSED,NSTEP,NFE,NJE
      COMMON /PCOM/ COEF1,COEF2,NM1
      C = 25.
      N = 20
      T0 = 0.
      H0 = 1.E-3
      DO 10 I = 1,N
 10     Y0(I) = 0.
      EPS = 1.E-3
      IERROR = 3
      MF = 10
      DX = 1./FLOAT(N)
      COEF1 = 1./DX**2
      COEF2 = .5*C/DX
      NM1 = N - 1
      WRITE(3,20)C,N,MF,EPS
 20   FORMAT(//49H EXAMPLE PROBLEM 3.. DIFFUSION-CONVECTION PROBLEM//
     1      4H C =,F6.1,6H   N =,I3//5H MF =,I3,8H   EPS =,E10.1///
     2      4H    T,6X,5HNSTEP,3X,3HNFE,3X,3HNJE,2X,2HNQ,4X,1HH,10X,
     3      4HY(1),9X,6HY(N/2),7X,4HY(N)//)
      NH = N/2
      NOUT = 10
      TOUT = .1/C
      INDEX = 1
      DO 60 IOUT = 1,NOUT
        CALL DRIVE(N,T0,H0,Y0,TOUT,EPS,IERROR,MF,INDEX)
        WRITE(3,40)T0,NSTEP,NFE,NJE,NQUSED,H0,Y0(1),Y0(NH),Y0(N)
 40     FORMAT(E9.1,3I6,I4,E10.2,3E12.4)
        IF (INDEX .NE. 0) GO TO 90
 60     TOUT = TOUT + .1/C
      CALL EXIT
 90   WRITE(3,100)INDEX
 100  FORMAT(///21H ERROR HALT   INDEX =,I3)
      CALL EXIT
      END
```

```
      SUBROUTINE DIFFUN(N,T,Y,YDOT)
      DIMENSION Y(1),YDOT(1)
      COMMON /PCOM/ COEF1,COEF2,NM1
      YDOT(1) = COEF1*(1. - 2.*Y(1) + Y(2)) - COEF2*(Y(2) - 1.)
      YDOT(N) = 2.*COEF1*(Y(NM1) - Y(N))
      DO 10 I = 2,NM1
        YDOT(I) = COEF1*(Y(I+1) - 2.*Y(I) + Y(I-1))
     1      - COEF2*(Y(I+1) - Y(I-1))
 10     CONTINUE
      RETURN
      END


      SUBROUTINE PEDERV(N,T,Y,PD,NO)
      RETURN
      END
```

The output of this code is as follows:

```
EXAMPLE PROBLEM 3.. DIFFUSION-CONVECTION PROBLEM

C =  25.0    N = 20

MF = 10    EPS =    1.0E-03


  T       NSTEP   NFE   NJE  NQ    H           Y(1)          Y(N/2)         Y(N)


---  MESSAGE FROM SUBROUTINE DRIVE IN EPISODE, THE O.D.E. SOLVER.  ---


KFLAG = -3 FROM INTEGRATOR AT T =     0.
 CORRECTOR CONVERGENCE COULD NOT BE ACHIEVED
 H HAS BEEN REDUCED TO      1.00000000E-04   AND STEP WILL BE RETRIED

 4.0E-03    12    27     0   3  7.19E-04  8.6400E-01  2.6398E-04  1.6973E-14
 8.0E-03    21    47     0   2  4.04E-04  9.6579E-01  2.1785E-02  3.4125E-07
 1.2E-02    26    57     0   3  1.08E-03  9.8924E-01  1.3763E-01  7.6861E-05
 1.6E-02    35    79     0   4  4.27E-04  9.9620E-01  3.4797E-01  1.8335E-03
 2.0E-02    44    92     0   4  4.27E-04  9.9855E-01  5.6965E-01  1.3616E-02
 2.4E-02    54   106     0   4  4.27E-04  9.9943E-01  7.3781E-01  5.9737E-02
 2.8E-02    63   120     0   4  4.27E-04  9.9977E-01  8.5155E-01  1.5310E-01
 3.2E-02    72   133     0   4  4.27E-04  9.9990E-01  9.1944E-01  2.8474E-01
 3.6E-02    82   147     0   4  4.27E-04  9.9996E-01  9.5797E-01  4.2863E-01
 4.0E-02    91   159     0   3  6.02E-04  9.9998E-01  9.7842E-01  5.8779E-01
```

Notice that a message was printed showing that on the very first step, the package could not pass the corrector convergence test without reducing the step size. It did this, automatically, and had no trouble thereafter. In any subsequent runs with this or similar problems, HØ should therefore be reset, say to 10^{-4}, in order to avoid this trouble.

To show the solution more clearly, a second run was made with calls to plotting routines inserted, plotting u vs. x for selected t values. The results of that run are shown in Figure 1.

Since the values of y are all between 0 and 1, the error option IERROR = 3 causes errors to be measured in the absolute sense here, and choosing IERROR = 1 would have given the same results. As before, we can estimate the global errors in these results by rerunning with a smaller EPS. This was done, and shows errors of about EPS = 10^{-3} or less.

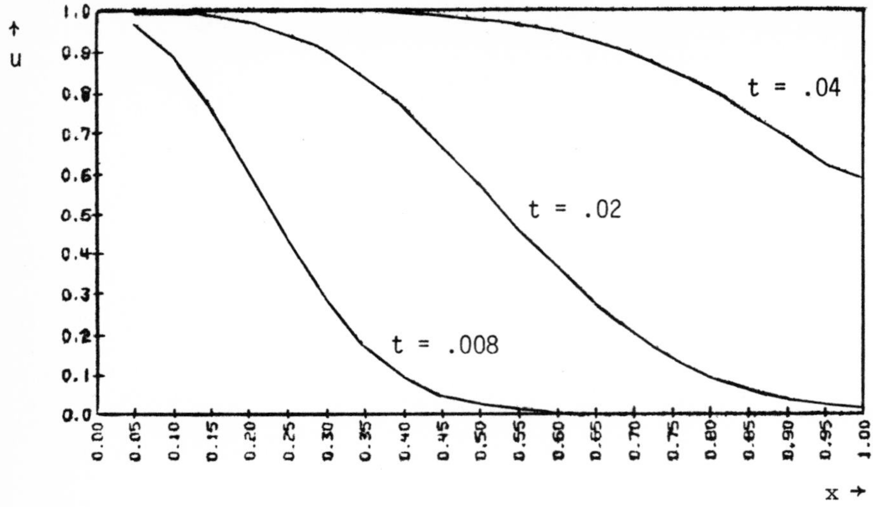

Figure 1. Output for Problem 3.

The performance of MF = 10 here appears satisfactory, considering the nature of the traveling wavefront and the values of N and c. The approximate time required for the front to travel over one spatial subinterval is $N^{-1}/c = .002$, and the average time step used is about one-fifth of this. However, this is not always the case. It is known [8] that in the limit of small c and large N, this problem is stiff, with a smallest time constant on the order of N^{-2}. In that case, we would choose MF = 21.

For the sake of illustration, we consider rerunning this problem with MF = 11 and 21, i.e., with a user-supplied Jacobian, and both the Adams and BDF formulas. To do this, we need a correct version of Subroutine PEDERV, such as the following:

```
      SUBROUTINE PEDERV(N,T,Y,PD,NO)
      DIMENSION Y(1),PD(NO,1)
      COMMON /PCOM/ COEF1,COEF2,NM1
      DO 10 J = 1,N
      DO 10 I = 1,N
10      PD(I,J) = 0.
      DO 20 I = 2,NM1
        PD(I,I) = -2.*COEF1
        PD(I,I-1) = COEF1 + COEF2
20      PD(I,I+1) = COEF1 - COEF2
      PD(1,1) = -2.*COEF1
      PD(1,2) = COEF1 - COEF2
      PD(N,N) = -2.*COEF1
      PD(N,NM1) = 2.*COEF1
      RETURN
      END
```

The outputs for MF = 11 and 21, respectively are as follows. All other input is unchanged except for HØ = 10^{-4}.

EXAMPLE PROBLEM 3.. DIFFUSION-CONVECTION PROBLEM

C = 25.0 N = 20

MF = 11 EPS = 1.0E-03

T	NSTEP	NFE	NJE	NQ	H	Y(1)	Y(N/2)	Y(N)
4.0E-03	11	20	4	4	8.54E-04	8.6497E-01	2.6369E-04	3.4420E-11
8.0E-03	15	25	5	4	1.48E-03	9.6617E-01	2.1566E-02	4.5063E-07
1.2E-02	18	29	6	4	1.48E-03	9.8925E-01	1.3857E-01	7.9843E-05
1.6E-02	20	32	7	4	1.95E-03	9.9619E-01	3.4835E-01	1.8565E-03
2.0E-02	22	34	7	4	1.95E-03	9.9858E-01	5.6704E-01	1.4476E-02
2.4E-02	24	38	8	4	2.62E-03	9.9939E-01	7.3811E-01	5.7465E-02
2.8E-02	25	39	8	4	2.62E-03	9.9975E-01	8.5158E-01	1.4696E-01
3.2E-02	27	41	8	4	2.62E-03	9 3992E-01	9.1964E-01	2.7937E-01
3.6E-02	28	43	8	4	2.62E-03	1.0000E+00	9.5789E-01	4.3312E-01
4.0E-02	30	47	8	4	2.62E-03	9.9997E-01	9.7842E-01	5.8280E-01

EXAMPLE PROBLEM 3.. DIFFUSION-CONVECTION PROBLEM

C = 25.0 N = 20

MF = 21 EPS = 1.0E-03

T	NSTEP	NFE	NJE	NQ	H	Y(1)	Y(N/2)	Y(N)
4.0E-03	13	24	4	3	5.81E-04	8.6347E-01	3.1148E-04	1.1695E-10
8.0E-03	18	31	6	3	1.14E-03	9.6538E-01	2.1516E-02	9.4179E-07
1.2E-02	21	37	6	3	1.14E-03	9.8887E-01	1.3730E-01	1.0940E-04
1.6E-02	25	41	7	3	1.55E-03	9.9606E-01	3.4833E-01	2.0338E-03
2.0E-02	27	44	7	3	1.55E-03	9.9850E-01	5.6862E-01	1.4716E-02
2.4E-02	30	47	7	3	1.55E-03	9.9940E-01	7.3967E-01	5.7009E-02
2.8E-02	32	49	7	4	2.19E-03	9.9975E-01	8.5222E-01	1.4566E-01
3.2E-02	34	52	9	4	3.84E-03	9.9990E-01	9.1948E-01	2.7866E-01
3.6E-02	35	54	9	4	3.84E-03	9 3996E-01	9.5744E-01	4.3434E-01
4.0E-02	36	56	9	4	3.84E-03	9.9999E-01	9.7826E-01	5.8541E-01

We see that in both cases, the code required about one-third as many steps or f evaluations as for MF = 10. However, in return, it required the extra computation associated with the Jacobian matrix, not required for MF = 10. To each evaluation of J, we can assign an approximate cost of 3 evaluations of f, as there are about 3N nonzero elements in J. If we now adjust NFE by the addition of 3*NJE, we find that the total costs for MF = 11 and 21 were 71 and 83 evaluations, respectively. This is considerably less than the cost for MF = 10. In fact, on this basis, MF = 11 is the least costly of all eight values of MF, for this problem, at the error tolerance chosen. However, the cost comparison made here is a rather crude one, in that it neglects the overhead associated with linear system solution, but probably overestimates the cost of J evaluation. Nevertheless, this example does show the need to be able to choose METH and MITER independently. This capability is not present, for example, in DIFSUB.

One final comment should be made regarding this problem, and regarding the use of the method of lines for the solution of partial differential equations generally. Solution by use of EPISODE, or any other stiff system solver, will generally require the capability of solving large linear systems which have a fairly well-defined sparse form, e.g., tridiagonal form in the example here. It is usually possible to take advantage of this matrix structure, rather than assume no particular structure as is done in EPISODE. Variants of both EPISODE and GEAR have been developed with this in mind. Thus, for example, the variants EPISODEB (unpublished) and GEARB [6] have been written to treat the matrix problem in band form. These codes substitute band matrix routines for full matrix routines, with the rest of the package virtually unchanged. For problems of this type, these variants offer a considerable savings in both run time and storage.

3. Package Description

In this section we give a few details on the EPISODE package as such, and discuss its availability. Complete details can be found in [1] and [4].

The EPISODE package consists of eight FORTRAN subroutines, to be combined with the user's calling program and Subroutines DIFFUN and PEDERV. Figure 2 shows the relationship of these routines. As discussed earlier, only Subroutine DRIVE is called by the user; the others are called within the package. In Figure 2, a downward sloping line from one box to another indicates that the lower routine is called by the upper one.

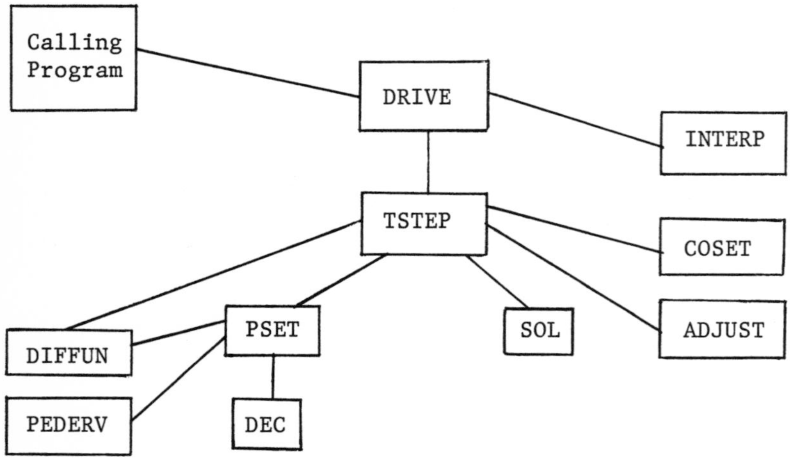

Figure 2. Structure of the EPISODE Package.

The functions of the eight package routines can be briefly summarized as follows:

- DRIVE sets up storage, makes calls to the core integrator, TSTEP, checks for and deals with error returns, and prints error messages as needed.
- INTERP computes interpolated values of y(t) at the user-specified output points, using an array of multistep history data.
- TSTEP performs a single step of the integration, and does the control of local error (which entails selection of the step size and order) for that step.
- COSET sets coefficients that are used by TSTEP, both for the basic integration step and for error control.
- ADJUST adjusts the history array when the order is reduced.
- PSET is called if MITER is 1 or 2. It sets up the matrix $P = I - h\beta_0 J$, where I is the identity matrix, h is the step size, β_0 is a scalar related to the method, and J is the Jacobian matrix. It then processes P for subsequent solution of linear algebraic systems with P as coefficient matrix.
- DEC performs an LU (lower-upper triangular) decomposition of an $N \times N$ matrix.
- SOL solves linear algebraic systems for which the matrix was factored by DEC.

The package is intentionally organized in a modular form in this manner in order to facilitate both the understanding of it and the modification of it as needed.

EPISODE has been written in a manner intended to maximize its portability among various computer installations. First of all, the language used is almost entirely ANSI Standard FORTRAN. Secondly, a new approach is taken toward the problem of single vs. double precision. As supplied outside LLL, the package is written in double precision. However, for all statements affected by the multiplicity of precision to be used, the single precision version of the statements are also given, and are imbedded in comment cards. Special flags are used to demark the two types of statements. We refer to this programming style as dual precision coding. In order to generate the single precision version of the package, one simply applies to it a converter subroutine, supplied with the package and also written in standard FORTRAN, which reads the special flags and makes the appropriate changes. This process is reversible, and the version of the converter needed for the reverse process can be obtained by applying the given converter routine to itself.

The "experimental" nature of EPISODE requires some explanation. First of all, the program is relatively new and has not been used extensively (in contrast to GEAR, for example), and so its position in the spectrum of existing available ordinary differential equation software is not yet clear. Secondly, testing has shown that, for some types of problems, the program spends more time on the linear system aspect of the algorithm than we feel it should. This behavior is related to the extent to which the matrix P changes during the solution of a problem, and we hope to study ways of improving the efficiency of the algorithm in this area. Finally, the dual precision style of programming has, as far as we know, never been used for a publicaly available code, and so is somewhat experimental.

The EPISODE package is being distributed by way of the following:

Argonne Code Center
Argonne National Laboratory
9700 South Cass Avenue
Argonne, IL 60439
Phone (312) 739-7711, Ext. 4366

It is available to anyone, worldwide. Requestors should send a blank magnetic tape, together with the character, channel, and density specifications, and ask for EPISODE, A.C.C. no. 675. A copy of the user's manual [4] will be returned with the tape. There is no charge for the first request from any given computer installation.

REFERENCES

[1] Byrne, G.D. and Hindmarsh, A.C., *A Polyalgorithm for the Numerical Solution of Ordinary Differential Equations, ACM Trans. Math. Software 1*, 71 (1975); also Lawrence Livermore Laboratory Report UCRL-75672 (1974).

[2] Gear, C.W., *Numerical Initial Value Problems in Ordinary Differential Equations* (Prentice-Hall, Englewood Cliffs, NJ, 1971).

[3] Gelinas, R.J. and Dickinson, R.P., *GENKIN-I: A General Kinetics Code — Atmospheric Applications*, Lawrence Livermore Laboratory Report UCID-16577 (1974).

[4] Hindmarsh, A.C., and Byrne, G.D., *EPISODE: An Experimental Package for the Integration of Systems of Ordinary Differential Equations*, Lawrence Livermore Laboratory Report UCID-30112 (1975).

[5] Hindmarsh, A.C., *GEAR: Ordinary Differential Equation System Solver*, Lawrence Livermore Laboratory Report UCID-30001, Rev. 3 (1974); program also available from Argonne Code Center.

[6] Hindmarsh, A.C., *GEARB: Solution of Ordinary Differential Equations Having Banded Jacobian*, Lawrence Livermore Laboratory Report UCID-30059, Rev. 1 (1975); program also available from Argonne Code Center.

[7] Lapidus, L., and Seinfeld, J.H., *Numerical Solution of Ordinary Differential Equations* (Academic Press, New York, 1971)

[8] Sincovec, R.F. and Madsen, N.K., *Software for Nonlinear Partial Differential Equations*, Lawrence Livermore Laboratory Report UCRL-75658, Rev. 1 (1974); also to appear in *ACM Trans. Math. Software 1* (1975).

SETKIN: A Chemical Kenetics Preprocessor Code

R. P. Dickinson, Jr. and R. J. Gelinas

1. INTRODUCTION

The purpose of this paper is to describe the internal structure of a chemical kinetics preprocessor code called SETKIN. This code is designed to translate a user-specified system of chemical rate equations into a system of chemical kinetic differential equations. We symbolically represent these differential equations as follows:

$$\frac{d\, y_i(t)}{dt} = f_i\left(y_1(t),\ \ldots,\ y_{NSP}(t),\ Q_1(t),\ \ldots,\ Q_{NRATE}(t)\right) + S_i(t) \quad . \tag{1}$$

In the above system, $y_i(t)$ symbolically represents the ith time varying chemical species concentration; NSP is equal to the number of chemical species in the problem; $Q_k(t)$ represents various time dependent rate coefficients; NRATE is equal to the number of rate processes; $S_i(t)$ represents a time varying source or sink term; and t represents time.

SETKIN produces an output file called CRACK1 which is used as input to one of a collection of programs designed to solve the chemical kinetics ordinary differential equation systems (ODE's). Frequently, the ODE systems which result from chemical problems have the property of stiffness, and it is because of this fact that each of our solver programs has incorporated into it a very powerful ODE solver called EPISODE [1]. The particular solver

---

Work performed under the auspices of the U.S. Energy Research & Development Administration. Contract #W-7405-ENG-48.

program a user specifies depends on the nature of the ODE system he is solving.

Perhaps the most complicated program used as a solver is called DRUNCH. This program is designed to solve atmospheric chemical kinetic problems with diurnally varying rate coefficients $Q_k(t)$. Part of the input for DRUNCH are such things as longitude, latitude, altitude, and time of year. Using a built-in calendar the program simulates day and night conditions for a period of time specified by the user. DRUNCH contains a subroutine called SOLRAD which calculates the time varying solar photodissociation rate coefficients ($Q_k(t)$).

Another solver in the collection is called CRUNCH. This program is used for time independent rate functions; that is, $Q_k(t)$ is a constant for each k and all t.

A new and very interesting solver is called SENSIT. Often a chemical kinetics system contains certain constants c_j which are supplied, say, by numerical experiment. Often these constants are not known with great accuracy. It is desirable to calculate changes in $y_i(t)$ due to changes, Δc_j, in c_j. Let c be one of the c_j, then SENSIT will solve a second system of ODE's along with the original system, and the solutions to the second system will be the functions $\partial y_i(t)/\partial c$. Using ideas from the calculus of variation we extend our sensitivity analysis even to time varying functions c, where as before the function is not known with full accuracy. A description of the program SENSIT can be found in [2].

This paper is considered a complement to the paper "GENKIN-I — A General Kinetics Code - Atmospheric Applications" [3]. GENKIN-I can be viewed as a user's manual to the chemical kinetics programs developed by the authors of this paper at the Lawrence Livermore Laboratory (LLL). This paper describes the internal structure of SETKIN which is at the heart of our kinetics package.

Although most of our work has centered around atmospheric problems, we feel that our programs have much greater flexibility. We feel that, with slight modifications, our procedures can be adapted to many other physical problems involving chemical kinetics. For purposes of this paper, we shall restrict ourselves to an example from atmospheric kinetics. However, it should be apparent to the reader how these procedures can be adopted in solving other types of problems.

Most of the following sections will be devoted to the description of important subroutines which are part of the GENKIN-I system. Two of these sections will describe key subroutines in SETKIN. Following sections will describe the differential equation function subroutine and also the associated subroutine which allows us to calculate the Jacobian of the ODE system. The Jacobian will be important to us from the standpoint of numerical accuracy, and also we will see that it is intimately related to the program SENSIT. In the final section we will more fully draw out this connection. Most of the following sections will be entitled by the actual subroutine name which has been used.

We include the following figure to summarize the GENKIN-I system. Each box represents either disk input or output files, or a program package within the kinetics system.

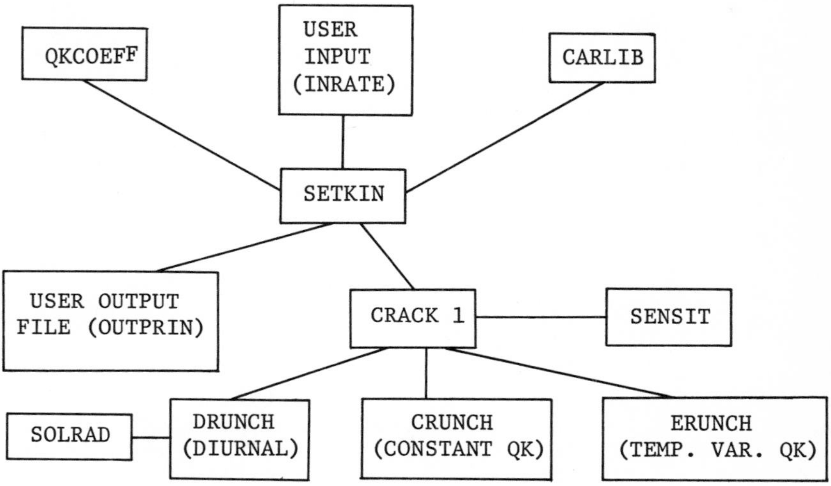

Figure 1.

2. EXAMPLE

The example which we consider consists of four reactions and three atmospheric species. It is the very simple, but classic, Chapman mechanism for atmospheric ozone kinetics:

$$O + O_2 \overset{M}{\rightarrow} O_3 \quad : \quad R88$$

$$O + O_3 \rightarrow 2O_2 \quad : \quad R89$$

$$O_2 \overset{h\nu}{\rightarrow} O + O \quad : \quad R93$$

$$O_3 \overset{h\nu}{\rightarrow} O + O_2 \quad : \quad R103 \text{ ,}$$

where the reaction library numbers are given by R88, R89, R93, and R103. In the following section we will discuss our reaction library. The symbols M and $h\nu$ which appear above the arrows mean that certain modifications will be made to the calculation of R88, R93, and R103. The modification due to M takes place because the reaction is in the presence of air acting as a third body. The modification due to $h\nu$ occurs from the presence of photons (radiation from the sun). These ideas will be discussed more fully in the next section.

3. QKCALC ($Q_k(t)$ calculation)

As part of the GENKIN-I system, the user is supplied with a catalogue entitled "LLL Atmospheric Data Library" [4]. This library contains data for most homogeneous gas phase reactions likely to play major roles in the stratosphere. It also contains a limited set of reactions designed for the modeling of tropospheric smog. The user chooses an appropriate subset of reactions from this library, and he supplies the library indexes to SETCRN. In the example we have chosen, the numbers are 88, **89**, 93 **and** 103.

SETCRN has access to a copy of reaction library, and this copy resides in a disk file called QKCOEFF (Q_k coefficients). We show a subset of QKCOEFF appropriate to the example below:

```
 5.1  E-12  Ø.       Ø.  NO2 + O = NO + O2                R8Ø
 1.1  E-12 -Ø.1      Ø.  NO2 + O + ZM = NO3 + ZM          R81
 1.23 E-13  Ø.    247Ø.  NO2 + O3 = NO3 + O2              R82
 Ø.                  Ø.  NO2 + ZRAD = NO + O              R83
 4.3  E-12  Ø.    385Ø.  NO3 + NO3 = 2NO2 + O2            R84
 Ø.         Ø.       Ø.  NO3 + ZRAD = NO2 + O             R85
 Ø.         Ø.       Ø.  NO2 + ZRAD = NO + O1D            R86
 3.8  E-3Ø -1.     17Ø.  O + O + ZM = O2 + ZM             R87
→1.Ø7 E-34  Ø.    -51Ø.  O + O2 + ZM = O3 + ZM            R88
→1.5  E-11  Ø.    23ØØ.  O + O3 = O2 + O2                 R89
 7.5  E-11  Ø.       Ø.  O1D + O2 = O + O21EGP            R9Ø
 5.   E-1Ø  Ø.       Ø.  O1D + O3 = O + O + O2            R91
 5.   E-1Ø  Ø.       Ø.  O1D + O3 = O2 + O2               R92
→Ø.         Ø.       Ø.  O2 + ZRAD = O + O                R93
 5.85 E-11  Ø.       Ø.  O1D + ZM = O + ZM                R94
 1.Ø7 E-34  Ø.    -51Ø.  O + O2 + N2 = O3 + N2            R95
 1.Ø7 E-34  Ø.    -51Ø.  O + O2 + O2 = O3 + O2 + O2       R96

 4.75 E-Ø5 -1.   39358.  O2 + ZM = O + O + ZM             R98
 4.5  E-11  Ø.    282Ø.  O21D + O3 = O + 2O2              R99
 5.   E-21  Ø.8      Ø.  O21D + ZM = O2 + ZM              R1ØØ
 6.   E-12  Ø.       Ø.  O21EGP + O3 = O + O2 + O2        R1Ø1
 1.6  E-15  Ø.       Ø.  O281EGP + ZM = O2 + ZM           R1Ø2
→Ø.         Ø.       Ø.  O3 + ZRAD(44Ø-85ØNM) = O + O2    R1Ø3

 Ø.         Ø.       Ø.  O3 + ZRAD(3ØØ-36ØNM) = O1D + O2  R1Ø5
 4.2  E-1Ø  Ø.     95Ø.  H + HO2 = HO + HO                R1Ø6
 3.6  E-11  Ø     295Ø.  H2O2 + O = HO + HO2 (1Ø7 + 13Ø)  R1Ø7
 Ø.         Ø.       Ø.  HNO3 + ZRAD = H + NO3            R1Ø8
```

Figure 2.

QKCOEFF is organized by card images, where each card contains exactly one reaction process. The reaction rate library number appears on the right-hand side of the card, and the number is the same as the actual card number in the disk file. Thus, it is very easy to read the specified card images given the user's reaction numbers. In the middle of each card is shown the reaction process. On the left side of each card is found three numbers which are used to calculate Q_k in many situations. In a following paragraph we explain this calculation more fully. It should be noticed, however, that this library is designed for atmospheric work, but there is nothing to stop the user from supplying a different library which meets the needs of his work. It should also be noticed that the symbol → has been replaced by the symbol =; we interpret = as →.

For many processes in the library, the rate coefficient of $Q_k(t)$ is calculated using the three numbers on the left of the card. We will call these numbers A, B, C. The standard formula which we then use for the rate coefficient is as follows:

$$QK = A * (TEMP^B) * EXP\ (-C/TEMP) ,$$

where TEMP stands for temperature and EXP stands for the exponential function with base e. If TEMP is time varying, then we pass the quantities A, B, C to the solver being used through the disk file CRACK1. Recall that CRACK1 is a data link between SETCRN and the selected solver.

The standard formula presented in the last paragraph is sometimes not completely appropriate, and several options are open to the user. Sometimes the standard formula has simple modifications as in the case of R88 (recall the M over the arrow). In these cases, special lists of indices are carried internally to SETKIN, and these lists are scanned to see if the user given reaction number is in the list. If this is so, the calculation of QK is modified in the correct way. In the following paragraphs we will explain these options.

Sometimes the standard formula or one of its modifications is not appropriate. First of all, it is possible for the user to completely override any calculations supplied by QKCALC. The user does this by reading cards with three pieces of information on each card: k, ireplace, RMULT. The variable k is a reaction library number for which override action is being called for. The variable ireplace is either 0 or an integer greater than 0. If ireplace = 0, then QK, which has been calculated, is scaled by RMULT; that is, the new QK equals the old QK times RMULT. If ireplace is greater than 0, then the new QK is RMULT. A second possibility is that QK might be a time varying function of a special nature, and in this case special subroutines must be added to the solver package for this calculation. In our diurnal codes, we have a special subroutine called SOLRAD in which some of the QK are time varying depending on radiation from the sun and the time of day. If a user wishes to hold a radiative rate fixed, he may use the override option. Special flags are carried through CRACK1 so that the solver will by-pass the time varying calculation.

Just as special functions have been written to handle time varying rate coefficients, so in the same way do we handle the sink or source terms $S_i(t)$ shown in (1). Thus, again we show how this program can be adapted to other kinetics work besides atmospheric work. We hope that we have developed a useful tool which can be modified to solve many chemical systems.

Two modifications in QKCALC will now be mentioned. If the reaction number appears in the list iadj (iadj is the name of an array containing certain library indexes), then the new QK is just the old QK times ZM (ZM is the concentration of air, a third body, present in the model). If the reaction number appears in the list ipress (pressure) then a similar modification will also occur.

Up to this point we have avoided the use of indexing. However, each QK will carry its own index. In our example, we would calculate QK(1), QK(2), QK(3), and QK(4); this means that NRATE for our problem is 4. The correspondence between indices and reaction library numbers is quite simple. The reaction library numbers are ordered with the smallest being first. Thus, $88 < 89 < 93 < 103$. The first element in the ordering corresponds to index 1, $1 \leftrightarrow 88$, and the second corresponds to index 2, and so on.

In our example, QK(1) is modified by multiplying by ZM, and QK(3) and QK(4) are both tagged as radiation rates. This means that SOLRAD will be used to calculate QK(3) and QK(4). In this case the information passed to CRACK1 would be NRATE = 4, QK(1), QK(2), and flags which state that QK(3) and QK(4) are radiative and will be calculated as functions of time.

4. BUSTER (Rate equation construction)

The subroutine BUSTER converts a user-specified set of reaction processes into a corresponding set of chemical kinetic differential equations as shown in (1). BUSTER uses a card library which is very similar to QKCOEFF called CARLIB (card library). We show a portion of CARLIB which we shall use for our example in the figure below.

It will be noted that the chemical equations shown in CARLIB are not completely identical to those shown in QKCOEFF; for example, R88 has the additional element ZM on both sides of the equal sign in QKCOEFF, and ZM is not present at all in R88 for CARLIB. Another example of this is R93 in which ZRAD appears in one library but not the other. These two examples are different, for in the last example, ZRAD does not appear on both sides of the equal sign. We shall explain the difference between these two libraries after we have examined the method of converting the rate processes into differential equations. One may think of the reaction process as shown in QKCOEFF as being complete. The process as shown in CARLIB is a device which is designed for the purposes of BUSTER.

```
NO2 + O = NO + O2 : R8Ø
NO2 + O = NO3 : R81
NO2 + O3 = NO3 + O2 : R82
NO2 = NO + O : R83
NO3 + NO3 = NO2 + NO2 + O2 : R84
NO3 = NO2 + O : R85
NO2 = NO + O1D : R86
O + O = O2 : R87
O + O2 = O3 : R88
O + O3 = 2O2 : R89
O1D + O2 = O + O21EGP : R9Ø
O1D + O3 = 2O + O2 : R91
O1D + O3 = 2O2 : R92
O2 = O + O : R93
O1D = O : R94
O + O2 + N2 = O3 + N2 : R95
O + O2 + O2 = O3 + O2 : R96

O2 = O + O : R98
O21D + O3 = O + 2O2 : R99
O21D = O2 : R1ØØ
O21EGP + O3 = O + 2O2 : R1Ø1
O281EGP = O2 : R1Ø2
O3 = O + O2 : R1Ø3

O3 = O1D + O2 : R1Ø5
H + HO2 = HO + HO : R1Ø6
H2O2 + O = HO + HO2 : R1Ø7
HNO3 = H + NO3 : R1Ø8
```

Figure 3.

In order to describe how BUSTER works, we show the differential equations which are produced by the example. An auxillary array, QQ, will make the description easier.

$$QQ(1) = QK(1) * O * O2$$

$$QQ(2) = QK(2) * O * O3$$

$$QQ(3) = QK(3) * O2$$

$$QQ(4) = QK(4) * O3$$

$$\dot{O} = 2*QQ(3) + QQ(4) - QQ(1) - QQ(2)$$

$$\dot{O}2 = 2*QQ(2) + QQ(4) - QQ(1) - QQ(3)$$

$$\dot{O}3 = QQ(1) - QQ(2) - QQ(4)$$

To reach the differential equations produced by BUSTER, certain rules are established in connection with the CARLIB library.

Rule 1: For each card in CARLIB, we think of elements to the left of the equal sign as loss or minus terms, and we think of elements to the right of the equal sign as production or plus terms. Thus, for R88, 0 and 02 are loss terms and 03 is a production term.

Rule 2: The numerical coefficients which appear in front of a chemical specie must be a positive integer. If an integer does not appear, it is understood that the coefficient associated with the specie is 1. Notice that an element may be repeated in an equation in more than one way. For example, in R89, 02 has a numerical coefficient of 2, whereas in R93, 0 is repeated twice. To BUSTER 0 + 0 and 20 are equivalent. Either style of presentation can occur in CARLIB.

Rule 3: The auxillary array QQ is calculated using the left hand or minus terms of each card from CARLIB. QQ(1) is calculated from the first card called for, namely R88. QQ(2) is calculated from R89; QQ(3) is calculated from R93, and QQ(4) comes from R1Ø3. Each QQ is formed as a product of a QK and the chemical species which appear on the left of the equal sign. If a chemical specie occurs more than once in the minus terms, then that specie will occur as a power in the product which makes up QQ, and the exponent of the power will be exactly the number of times the specie occurred as a minus term. Occurring more than once is equivalent to having numerical coefficients greater than 1. As an example, let us assume that species y_1, y_2, y_3 appear as minus terms for some card of CARLIB, and let us assume their numerical coefficients are m_1, m_2, m_3. Then

$$QQ = QK * y_1^{m_1} * y_2^{m_2} * y_3^{m_3} .$$

In our example, 0 and 02 occur in minus terms for R88, and they occur with numerical coefficients 1. Thus, QQ(1) = QK(1) * 0 * 02. It is noticed that we produce one QQ and one QK for each reaction process specified.

Rule 4: Whereas each selected card from CARLIB determines a QQ term, one must look at all the selected cards to determine a differential equation for a particular species. The differential equation which we develop will be a linear combination of the QQ's where the coefficients of the QQ's will be integers. The differential equation system will be developed cumulatively as we pass from one card to the next. In order to understand how to select the QQ's we look at a particular example.

Suppose we are developing a differential equation for 02. We associate with the ith card of the selected set from CARLIB a pair of integers (n_i,m_i), where n_i is the number of times 02 occurs on the minus side and m_i is the number of times 02 occurs on the plus side of the ith card. Equivalently, n_i is the numerical coefficient of 02 on the minus side of the ith card. Each card will contribute one and only one term to the differential equation of the selected specie. In particular the ith card will contribute the term: $(m_i-n_i) * QQ(i)$.

Notice that if $m_i = n_i$ the coefficient for QQ(i) will be zero, or QQ(i) will not be present as a term in the differential equation for O2. Also notice that the coefficient of QQ(i) can be a positive or negative integer, depending on the relative sizes of m_i and n_i.

For the example of this paper, we develop the differential equation for O2. In this case $(n_1,m_1) = (1,0)$, $(n_2,m_2) = (0,2)$, $(n_3,m_3) = (1,0)$, $(n_4,m_4) = (0,1)$. We then have:

$$\dot{O}2 = (0-1)*QQ(1) + (2-0)*QQ(2) + (0-1)*QQ(3) + (1-0)*QQ(4) \quad ,$$

or

$$\dot{O}2 = -QQ(1) + 2*QQ(2) - QQ(3) + QQ(4) \ .$$

If we arrange the equation with plus terms first followed by negative terms, we have:

$$\dot{O}2 = 2*QQ(2) + QQ(4) - QQ(1) - QQ(3) \ ,$$

and it is in this form which we present the equation.

BUSTER operates on the given card set to produce the ODE system. It records the structure of this system in various arrays which it passes to a particular solver through the file CRACK1. We now describe these various arrays.

The array SØ(i) contains the name of the ith new species encountered in the scan of the cards. The number of elements in this array is NSP, the number of species in the problem. There will normally be NSP differential equations in the ODE system, but certain exceptions which we describe later can reduce that number. For our example, SØ(1) = O, SØ(2) = O2, SØ(3) = O3. The order in which species occur is the order in which they come from left to right scans and from card to card. They enter into the order as seen with first being first and last being last. The differential equations are presented in the same order as SØ.

The next two arrays describe the structure of the QQ's.

The array icon contains the number of species used in the product to form QQ's. Suppose QQ(1) (card 1) involves species y_1, y_2, y_3 on the minus side, with coefficients m_1, m_2, m_3. Then $icon(1) = m_1 + m_2 + m_3$. In our example, icon(1) = 2, icon(2) = 2, icon(3) = 1, and icon(4) = 1.

The array icon1(j,i), i = 1, NRATE and j = 1, icon(i), actually contains the encoded names of the species used to make up QQ's. By encoded name we mean that we associate with SØ(1) the number 1, or 1 ↔ O. Here we let i be fixed to generate QQ(i) while j runs from 1 to icon(i). For our example, icon1(1,1) = 1, icon1(2,1) = 2; this means that QQ(1) = QK(1) * O * O2; icon1(1,2) = 1, icon1(2,2) = 3; icon1(1,3) = 2; icon1(1,4) = 3. If an element had an exponent m_1, then that element's encoded name would occur m_1 times in icon1(,i). In other words, it would appear as a factor m_1 times in making up QQ.

The next three arrays describe the structure of the ODE system. For any particular species they tell the number of QQ terms in the ODE, the coefficients of each QQ, and which QQ's are present.

The array icon4(i), i = 1, NSP, contains the number of QQ's present in the ith differential equation. In our example, icon4 (1) = 4, icon4(2) = 4, icon4(3) = 3.

The array icon2(j,i), i = 1, NSP and j = 1, icon4(i), contains the names of encoded QQ's. To describe the ith equation i is held fixed while j runs. In our example,

icon2(1,1) = 3, icon2(2,1) = 4, icon2(3,1) = 1,
icon2(4,1) = 2, icon2(1,2) = 2, icon2(2,2) = 4
icon2(3,2) = 1, icon2(4,2) = 3, icon2(1,3) = 1,
icon2(2,3) = 2, icon2(3,3) = 4.

Finally, the last array coef(j,i) contains the numerical coefficients of the QQ's. As i is held fixed, we describe with running j the coefficients for the ith equation. In our example,

coef(1,1) = 2., coef(2,1) = 1.,
coef(3,1) =-1., coef(4,1) =-1.,
coef(1,2) = 2., coef(2,2) = 1.,
coef(3,2) =-1., coef(4,2) =-1.,
coef(1,3) = 1., coef(2,3) =-1.,
coef(3,3) =-1.

This completes the description of the arrays produced by BUSTER. Given these arrays, it is very easy to construct a differential equation function necessary for the solvers, and we will describe such a function and its modifications in the section entitled DIFFUN. Another easy calculation is finding the Jacobian of the ODE system, and this we also describe in a following section. However, before leaving, we address two points which were brought up during this section.

The first point involves differences between the two disk files QKCOEFF and CARLIB. Two examples were shown where these differences exist; for example, ZM occurs on both sides of the equal sign in R88. The effect of this would be to incorporate ZM into the product of factors making up QQ(1), but also we would produce a differential equation for ZM in which no QQ's would be present. This is because $n_i = m_i$ for all cards we have selected from QKCOEFF, the more complete library. To avoid these difficulties, we removed ZM from both sides of the equal sign in CARLIB, and we added R88 to one of the special lists which were found in QKCALC, namely iadj. The last list had the property that ZM was worked into the standard QK as a factor. Notice that, by removing ZM from CARLIB for R88, no differential equation is produced. The natural equation is $\dot{ZM} = 0$, which states that ZM is a constant for the problem. Our method avoided ZM altogether except as a factor in QK. The second example for R93 works in the same way. ZRAD is to be included as a factor (due to photons) into a QK and held fixed or varied by a special subroutine in a diurnal solver. However, we did not want a differential equation developed for ZRAD. It is seen that these modifications are devices which can make the code somewhat more flexible, but they may be looked at as artificial.

As a second point, it was mentioned that sometimes there are fewer differential equations than species. This is merely an option we have supplied the user. If he chooses to hold certain species fixed during a run, then we delete from the ODE system, differential equations for those species. This is done by a subroutine called SHUFFLE. SHUFFLE rearranges the SØ list so that the names of deleted species end up on the bottom of the list; it also reencodes the necessary icon structure arrays.

The final sections of this paper discuss how to use the structure arrays from BUSTER in various applications.

5. DIFFUN (Differential Equation Function)

This subroutine takes as input arguments NEQN, T, Z, DZ, where NEQN is the number of differential equations; and, as we have seen, NEQN $\leq$ NSP, T stands for time, Z is a vector whose ith component is the concentration of species i at time T, and DZ is a vector whose ith component is the time derivative of the ith specie at time T. Given the structure arrays from BUSTER, it will be seen that this calculation of DZ by DIFFUN is fairly easy. This subroutine is used internally to the ODE system solver EPISODE.

We now outline the structure of this calculation.

Step 1: (Calculation of QQ's)

```
      Do   1   i=1,NRATE
      j1 = icon(i)
      QT = QK(i)
           Do   2   j=1,j1
           k = icon1 (j,i)
 2         QT = QT * Z(k)
 1    QQ(i) = QT
```

This part of the subroutine then consists of two simple nested do loops. The inner loop builds QT up as a product of specie concentrations and QK(i). In the case of time varying QK, we would first call a special subroutine to calculate QK at the given T. In our diurnal code this routine is called SOLRAD.

Step 2: (Calculation of DZ)

This part of the subroutine will be exactly analogous to Step 1, but instead of calculating a running product, we calculate a running sum.

```
      Do   3   i=1,NEQN
      SUM=Ø.
      j1 = icon4(i)
           Do   4   j=1,j1
           k = icon2 (j,i)
 4         SUM = SUM + COEF(j,i) * QQ(k)
 3    DZ(i) = SUM
```

If any source or sink terms are present as in (1), then at this point they should be evaluated and added to DZ.

6. PEDERV: (Jacobian Calculation)

Given an ODE system with NEQN equations such as (1), it is numerically useful to calculate a matrix J called the Jacobian of the system. We shall also see in a later section that J will be useful for doing sensitivity analysis. The matrix J is of order NEQN×NEQN and the (i,j)th element is given by $\partial f_i/\partial y_j$. We note that J is independent of any functions of t, in particular source and sink terms $S_i(t)$. By setting a flag in EPISODE, the ODE solver will use PEDERV as part of its algorithm.

Input and output to PEDERV consists of NEQN, T, Z, PD, where the first three arguments are the same as for DIFFUN, and where PD is an NEQN×NEQN matrix in which we return the Jacobian.

Let us analyze the structure of J.

$$f_i = \sum_{k=1}^{\text{icon4}(i)} \text{coef}(k,i)*\text{QQ}\left(\text{icon2}(k,i)\right) \quad ,$$

where the QQ terms are products of y's in our earlier notation (and z's presently). From this we see that

$$\frac{\partial f_i}{\partial y_j} = \sum_{k=1}^{\text{icon4}(i)} \text{coef}(k,i)* \frac{\partial \text{QQ}(\text{icon2}(k,i))}{\partial y_j} \quad . \tag{2}$$

Thus, to form J we must simply form partial derivatives of the QQ's, and then go through exactly the same summation procedure that was used in DIFFUN.

To accomplish this calculation, we use a temporary array called QDOT. This array is a vector of length at least NRATE, and the ith component of QDOT will be $\partial QQ(i)/\partial y_j$. Letting j be fixed, QDOT is formed, and then using (2), we form a column of J. We next describe the formation of QDOT.

The first step is to scan QQ(i) for the presence of y_j as a factor. It will also be important to know the multiplicity of this factor. If y_j is not present, then QDOT(i) = 0. We show sample FORTRAN to accomplish this scan.

```
      Do   2   k=1, icon(i)
      if (iconl(k,i)-j)2, 1, 2
 1    isav = k
      GO TO 3
 2    CONTINUE
      QDOT(i) = Ø.
 3    CONTINUE
```

Recall that iconl(k,i) contains the names of species used in forming QQ(i). If the loop 2 is completed, then y_j is not a factor of QQ(i) and QDOT(i) = Ø. If the factor is found, isav contains the index of the first location in the list where this factor occurs. Suppose that QQ(i) = QK(i) * O * O2 * O2; here O2 occurs with multiplicity 2, and if y_j = O2, then isav = 2 since the first occurrence of O2 is spot 2 in the list.

If we go to 3 in the above scan, then we proceed as follows. Our next move is to form QQ as before, but we modify in two ways. First, we form QQ as before but we delete one y_j from the product. The factor which we delete is tagged by isav. Secondly, we scan the list and determine the multiplicity; then QDOT(i) = mj * modified QQ(i). All we have described is the natural way of forming the derivative. It is not hard to construct such coding using our structure lists.

7. SENSIT: (Sensitivity Analysis)

Associated with systems of type (1) one will often find constants which enter the structure. In the examples we are considering, two possibilities occur: first, initial conditions of the ODE system, and second, constant QK's in the system. We would like to appraise the sensitivity of our system to such constants. In other words, we would like to predict changes in y given small changes in the constants. Using the tools which we have developed, it is easy to build a program which does this type of calculation (at LLL we call this program SENSIT).

Although there may be many constants in our system, let us for the moment think of only one constant c as being variable. Our ODE system is written in the following form:

$$\dot{y}_i = f_i\ (y_1, y_2, \ldots, y_n, t, c) \qquad i = 1, \ldots, N\ .$$

We now introduce new variables Z_i given by $Z_i = \partial y_i/\partial c$. Using certain theorems from calculus, we develop N differential equations whose solutions are the Z_i. This second system will be solved simultaneously with the first system, again using EPISODE.

To develop the differential equations, let us differentiate the Z_i with respect to t.

$$\frac{\partial z_i}{\partial t} = \frac{\partial}{\partial t}\frac{\partial y_i}{\partial c} = \frac{\partial}{\partial c}\frac{\partial y_i}{\partial t} = \frac{\partial}{\partial c} f_i\ .$$

Simple use has been made on a theorem on interchanging the order of differentiation. We now think of f_i as a function of all the y_i of the system, and of each y_i as a function of c and t, $y_i(c,t)$. Thus, to differentiate f_i with respect to c, we must look for the explicit variable and also the implicit presence of c. The derivative, dz_i/dt, is developed as follows:

$$\frac{\partial z_i}{\partial t} = \dot{z}_i = \frac{\partial f_i}{\partial c} + \sum_{j=1}^{N} \frac{\partial f_i}{\partial y_j}\frac{\partial y_j}{\partial c}\ , \qquad \text{or}$$

$$\dot{z}_i = \frac{\partial f_i}{\partial c} + \sum_{j=1}^{N} \frac{\partial f_i}{\partial y_j} z_j \qquad i = 1, \ldots, N \quad .$$

Using matrix notation, we rewrite this equation as

$$\dot{z} = f_c + Jz \quad ,$$

where $\dot{z}$ and z are N-vectors, we recognize J as the Jacobian of our original system, and f_c is a vector of length N, where the ith component of f_c is $\partial f_i/\partial c$, the explicit differentiation of f_i with respect to c. If c does not appear explicitly in any f_i, then we have $\dot{z} = Jz$. This will be the case if c is the initial value of some y_i.

At this point, it is fairly obvious that many of the necessary tools are present to build SENSIT. The Jacobian matrix is present through PEDERV. It remains to make simple modifications to DIFFUN in order to solve the system of size 2*N. One modification involves a call to PEDERV, and another modification involves calculating f_c. This last calculation will be very similar to calculations in the Jacobian routine. As we have mentioned before, [2] contains information on SENSIT and its applications.

REFERENCES

[1] A.C. Hindmarsh and G.D. Byrne, EPISODE: An Experimental Package for the Integration of Systems of Ordinary Differential Equations, UCID-30112, LLL, May, 1975.

[2] R.P. Dickinson and R.J. Gelinas, Sensitivity Analysis of Ordinary Differential Equation Systems — A Direct Variational Method, UCRL-76811, LLL, March, 1975.

[3] R.P. Dickinson and R.J. Gelinas, GENKIN-I, A General Atmospheric Code for Atmospheric Applications, LLL, UCID-16577, August, 1974.

[4] W.H. Duewer and R.J. Gelinas, LLL Atmospheric Data Library, Edition 4, UCID-16206, LLL, January, 1973.

Numerical Methods for Mass Action Kinetics

Lennart Edsberg

1. INTRODUCTION

One of the simplest reactor models occurring in chemical technology is the isothermal batch reactor operating under constant volume. The simplicity is not only a consequence of the idealized conditions under which the reactor is supposed to be working but also that the mathematical description becomes simple at least formally.

The physical and chemical conditions under which this model is valid is that a number of reactions, say n, simultaneously take place between m species in a physical environment where the contributions from diffusion and convection are neglected, only the chemical kinetics contribute to the rate terms. Assume that the chemical species G_i, i=1,2,...,m react in n elementary reactions

$$\sum_{i=1}^{m} p_{ij} G_i \rightarrow \sum_{i=1}^{m} q_{ij} G_i \qquad j=1,2,\ldots,n \qquad (1.1)$$

where the matrices P and Q with entries p_{ij} and q_{ij} respectively are the reactant and product matrices respectively. Introduce the stoichiometric matrix

$$S = P - Q \qquad (1.2)$$

and the concentrations c_i, i=1,2,...,m of the species and the system can be described by the following set of ordinary differential equations (ODE's), see e.g. (Gavalas 1968)

$$\dot{\bar{c}} = S\,\bar{r}(\bar{c}) \qquad (1.3)$$

where $\bar{r}^T = (r_1, r_2, \ldots, r_n)$ is a vector of rate functions r_i which depend on intensive thermodynamical variables of which temperature T and concentrations c_i (moles/unit volume) are the most important.

(1.3) is a simple consequence of mass conservation of the species G_i. The big problem is to obtain analytical expressions for the rate functions r_i. Several theories have developed for the prediction of the rates of elementary reactions. Collision theory in ideal gases and transition state theory are examples, cf Banford, Tipper (1969). A rule which is used very often is the mass action law which states that the reaction rates are

$$r_j = K_j \prod_{k=1}^{m} c_k^{p_{kj}} \qquad (1.4)$$

where K_j is some temperature dependent constant, the rate constant.

Although sometimes criticized (see e.g. Oster and Perelson (1974)) for its lack of consistency with thermodynamics the mass action law is one of the most frequently used "law" in practical engineering as well as in chemical mechanism models.

The very restrictive physical and chemical assumptions adhered to the model (1.3) imply that this model very seldom can be used in practise. However, with just a few additional modifications models are obtained which do occur frequently in practise.

A. If we allow continuous inlet and outflow of species (still under the assumption of constant reactor volume) the model is (see e.g. Aris 1965)

$$\dot{\bar{c}} = \frac{1}{\theta}(\bar{c}_f - \bar{c}) + S\,\bar{r}(\bar{c}) \qquad (1.5)$$

which is the well-known CSTR (Continuous-flow Stirred Tank Reactor). Here θ is the holding time and $\bar{c}_f^T = (c_{1f}, c_{2f}, \ldots, c_{mf})$, c_{if} being the concentration of G_i in the inflow. Under certain circumstances the CSTR can be treated with the same formalism as the batch reactor, see Feinberg and Horn (1974).

B. If we cannot neglect the temperature variations due to the reactions taking place we must add an ODE for the energy balance

$$\dot{T} = \frac{1}{\theta}(T_f - T) + \Delta\bar{H}^T\bar{r} + Q(T) \qquad (1.6)$$

where T_f is the inlet temperature, ΔH_i is the heat of reaction nr i, $Q(T)$ is some function describing the cooling or heating of the reactor. The temperature is coupled to (1.3) nonlinearly since K_j in (1.4) depends on T in a way which is often modelled by Arrhenius' law

$$K_j = k_j \exp(-E_j/RT) \qquad (1.7)$$

where k_j is the so called frequency factor, E_j the activation energy and R the universal gas constant.

C. The steady state of the PFTR (Plug Flow Tubular Reactor) can be modelled by ODE's very similar to (1.3). If F_i is the molar flux (moles/unit area and unit time) of component G_i the ODE's are

$$\bar{F}' = S\,\bar{r}(\bar{c}) \qquad (1.8)$$

where ' denotes differentiation with respect to the length coordinate of the reactor. If the ideal gas law is assumed to be valid, we have the following coupling between c_i and F_i

$$\bar{c} = \frac{p}{RT} \cdot \frac{\bar{F}}{F} \quad \text{where } F = \sum F_i \tag{1.9}$$

see e.g. Aris (1965). For further references see Grover (1967) and Edsberg (1975).

If diffusion and/or convection cannot be neglected the models usually become much more difficult comprising nonlinear partial differential equations, see e.g. Himmelblau and Bishoff (1968).

From the models just described it is obvious that (1.3) plays a very important part in the mathematical treatment of chemical reactors. We will here treat the initial value problem, i.e.

$$\bar{c}(t_0) = \bar{c}_0 \tag{1.10}$$

is given together with (1.3).

In this paper we will treat the numerical solution of the initial value problem (1.3), (1.10). In many applications the corresponding ODE's will be stiff due to the wide spread in magnitude of the rate constants K_j. The reason for special investigation of these systems are twofold:

1. The right hand sides have a special structure which can be utilized in the numerical implementation of a stiff method. Efficiency can then be gained compared to using a general purpose method.

2. Certain numerical difficulties have been observed on some practical examples; general purpose programs have failed on some test examples.

2. SOME MATHEMATICAL PROPERTIES OF MASS ACTION KINETICS

Assuming the validity of the mass action law the ODE's can be constructed once the elementary reactions are given. The matrices P and Q are uniquely determined from the reaction network. It is convenient to introduce a special notation for (1.3) when (1.4) is used.

Definition 1. Let $\bar{x}$ be $n\times 1$, $\bar{y}$ be $m\times 1$, A $m\times n$. Then $\bar{y} = \bar{x}^A$ means

$$y_i = \prod_k x_k^{a_{ik}}$$

Definition 2. $dg\bar{x} = X$, where X is diagonal and $X_{ii} = x_i$.

(1.3), (1.4) and (1.10) can now be written

$$\dot{\bar{c}} = S\ K\ \bar{c}^{-P^T} \quad , \ \bar{c}(0) = \bar{c}_0 \tag{2.1}$$

where $K = \mathrm{dg}K_i$. (2.1) is in general set of nonlinear ODE's. In the special case every column of P consists of only one element = 1, which we have in first order kinetics (see Wei and Prater 1962), we have a linear system with constant coefficients (provided the temperature is constant).

Example 1. The well-known enzyme-substrate model

$$E + S \underset{k_{-1}}{\overset{k_1}{\rightleftharpoons}} C \overset{k_2}{\rightarrow} E + P$$

is described by

$$\begin{aligned}
\dot{x}_1 &= -k_1x_1x_2 + k_{-1}x_3 + k_2x_3 \\
\dot{x}_2 &= -k_1x_1x_2 + k_{-1}x_3 \\
\dot{x}_3 &= k_1x_1x_2 - k_{-1}x_3 - k_2x_3 \\
\dot{x}_4 &= k_2x_3
\end{aligned} \tag{2.2}$$

where x_1, x_2, x_3, x_4 are the concentrations of E,S,C and P respectively. For this system we thus have the matrices S and P as follows:

$$S = \begin{pmatrix} -1 & 1 & 1 \\ -1 & 1 & 0 \\ 1 & -1 & -1 \\ 0 & 0 & 1 \end{pmatrix} \quad P = \begin{pmatrix} 1 & 0 & 0 \\ 1 & 0 & 0 \\ 0 & 1 & 1 \\ 0 & 0 & 0 \end{pmatrix}$$

For a closed system, which is assumed when (2.1) is valid, the total mass of the reacting system is preserved. This is expressed by

$$\bar{m}^T\bar{c} = \bar{m}^T\bar{c}_0 = k_0 \tag{2.3}$$

where $m_i > 0$ is the molecular weight of G_i and k_0 is a constant > 0.

We have
$$\bar{m}^T\dot{\bar{c}} = 0 \tag{2.4}$$

Thus, $\dot{\bar{c}}$ belongs to a subspace L orthogonal to the vector $\bar{m}$. Since all components of $\bar{m}$ are positive the normal vector of L points into the positive orthant. $\bar{c}$ belongs to a hyperplane L_0

parallell with L and intersecting the positive coordinate axes at intercepts $z_i > 0$.

It is consequently obvious that once $\bar{c}_0$ is chosen k_0 is determined and L_0 also. Less obvious is that for a system of ODE's of type (2.1) which has been constructed from the mass action law the solution $c_i(t) \geq 0$ if $c_{i0} \geq 0$. This, however, can be proved, see Edsberg (1975B) and as a consequence all solutions with nonnegative initial values belong to an invariant manifold Ω_0, usually called the reaction simplex.

From numerical point of view it is convenient to have the dependent variables c_i scaled so that $0 \leq c_i \leq 1$ for all t. This can be achieved in the following way: Let

$$w_i = \frac{\min(m_i)}{\bar{m}^T \bar{c}_0} c_i = \frac{\min(m_i)}{\bar{m}^T \bar{c}} c_i \qquad (2.5)$$

Then we have

$$w_i \leq \min(m_i)/m_i \leq 1 \qquad (2.6)$$

Inserting (2.5) into (2.1) gives

$$\dot{\bar{w}} = S\, K' \,\bar{w}^{P^T} \quad ; \quad \bar{w}(t_0) = \bar{w}_0 \qquad (2.7)$$

where $K' = q\, K\, (\mathrm{dg}\, \bar{q}^{P^T})^{-1}$, where $q = \min(m_i)/\bar{m}^T\bar{c}_0$ and $\bar{q} = (q,q,\ldots,q)$. Thus only K is affected by this scaling. However, if the components are scaled with different factors

$$\bar{w} = D\, \bar{c} \qquad (2.8)$$

where D is a diagonal matrix of scaling factors, S will also be changed.

The existence and uniqueness of an equilibrium point $\bar{c}_\infty$ to (2.1) has attracted much interest recently. Very thorough treatment of these and related questions have been made by Horn and Jackson (1972A), Horn (1972B), Feinberg (1972) and Feinberg and Horn (1974). They introduce a concept they call complex balancing (closely related to the principle of detailed balancing or microscopic reversibility, cf Wei (1962) and Horn and Jackson (1972A)), CB, which can be established a priori for a system (2.1) by examining rather simple algebraic conditions on the structure of the chemical reaction system, thus no analysis of the ODE's is necessary. When CB is fulfilled they have shown that this is a sufficient condition for the existence of a unique equilibrium point $\bar{c}_\infty$ lying in the interior of Ω_0. Thus for CB mass action systems all solutions starting in Ω_0 or on $\partial\Omega_0$ will eventually convergence to $\bar{c}_\infty$. However, this is not true for all mass action kinetics; certain problems of type (2.1) may exhibit bistability (several equilibrium points) or sustained oscillations, a fact which is of great use in the modelling of certain biochemical systems, where this situation may occur in experiments.

3. APPROXIMATE METHODS FOR MASS ACTION KINETICS

In an earlier paper by Edsberg (1974) a brief survey of some methods which has been used in order to solve (2.1) approximately was made. References to other review articles on the area were also given. A rough classification of these methods can be made:

A. Methods which treat (2.1) without any simplifying approximation like the steady-state approximation (see later). Since systems in chemical kinetics are very often stiff, explicit integration techniques are very inefficient so implicit methods should be used in order to avoid the small stepsizes. Several program packages especially designed for chemical kinetics problems have been developed by e.g. Garfinkel (1968) (explicit technique), Curtis and Chance (1974) and Edsberg (1975A) (implicit techniques).

B. Methods which are based on some initial approximation of the system (2.1) with subsequent numerical treatment of the approximated system. One such method is the well-known steady-state approximation which assumes that the fast (stiff) components are at equilibrium while the slow (non-stiff) components generate a smooth solution. Hence we have

$$\begin{aligned} \dot{\bar{c}}_1 &= \bar{f}_1(\bar{c}_1,\bar{c}_2) \\ \dot{\bar{c}}_2 &= \bar{f}_2(\bar{c}_1,\bar{c}_2) \end{aligned} \tag{3.1}$$

where $\bar{c}^T = (\bar{c}_1^T,\bar{c}_2^T)$ and $\bar{c}_1$ corresponds to the fast variables and $\bar{c}_2$ to the slow ones. The stiff components are approximately at equilibrium due to heavy cancellations between large terms in the right hand side of (3.1) so we can write

$$\begin{aligned} \varepsilon\dot{\bar{c}}_1 &= \bar{g}_1(\bar{c}_1,\bar{c}_2) \\ \dot{\bar{c}}_2 &= \bar{f}_2(\bar{c}_1,\bar{c}_2) \end{aligned} \tag{3.2}$$

where ε is a small parameter. The steady state approximation consists of letting $\varepsilon = 0$. This zeroth order approximation of the singularly perturbed system (3.2) will give a very poor representation of the transient. More accurate techniques based on asymptotic expansion of the solution of (3.2) in powers of ε to the first term have been developed by Aiken and Lapidus (1974). Earlier program packages based on the steady-state approximation coupled with a Runge-Kutta technique were developed by e.g. Snow (1966) and Lawrendeau and Sawyer (1970). Another technique developed by Clark and Groner (1972) is applicable when all reactions involved are reversible, i.e. (1.1) has a forward as well as a backward rate constant (and can then be written as two irreversible reactions of type (1.1)). In their approach separation is made between the slow and the fast reactions instead of slow and fast components. The numerical technique consists of an explicit method (of Runge-Kutta type) combined with a steepest descent and Newton technique to minimize the thermodynamic free energy function corresponding to the fast reactions in order to force them to equilibrium. Applied to the reaction system in example one (2.2) a steady state technique consists of letting $\dot{x}_3 = 0$ (usually known under the name Michaelis-Menten approximation) while a technique of the previously described type consists of letting the first reversible reaction to be in equilibrium the whole time (sometimes known as the Briggs-Haldane approximation, cf Aris (1969)).

The method to be described here is based on a stiff integration package as developed by Lindberg (1974). Utilizing certain characteristic features about mass action kinetics to be described later this method has been implemented as a program package for interactive simulation of systems of type (1.5), (1.6), i.e. the nonisothermal CSTR where the rate functions obey the mass action law, see Edsberg (1975A). The program is an extension of a previous version as described in Edsberg (1974). Lindberg's method is based on the implicit midpoint method with smoothing and extrapolation and it has turned out to be of an efficiency comparable to other good methods for stiff systems as e.g. Gear's method, cf the comparison of stiff methods by Enright, Hull and Lindberg (1975).

What will be discussed here is not the details of the program, but some special features and properties of mass action kinetics. These points of view are applicable to many other discretization methods, not only the implicit midpoint method.

3.1. The Jacobian evaluation

In an earlier paper (Edsberg (1974)) it is shown that the Jacobian is simple to evaluate for mass action kinetics. The matrix S, which is in general rank deficient, can be represented in the following way. Let Q be a permutation matrix which brings S on the form

$$Q\,S = \begin{pmatrix} S_1 \\ S_2 \end{pmatrix} \quad , \; S_2 = L\,S_1 \qquad (3.3)$$

Here S_1 has full rank r, so the rows of S_1 are linearly independent. Using (3.3) (2.1) can (after permutation) be brought on the form

$$\dot{\bar{c}}_1 = S_1\,K\,\bar{c}^{-P^T} \qquad \bar{c}_1(0) = \bar{c}_{10} \qquad (3.4)$$

and $\bar{c}_2$ which is needed in the evaluation of $\bar{c}$ is obtained from

$$\bar{c}_2 = \bar{c}_2(0) + L(\bar{c}_1 - \bar{c}_1(0)) \qquad (3.5)$$

The rank deficiency depends on mass conservation described in chapter 2 and conservation of atomic species. The reduction of the number of equations from m to r is of importance for the efficiency since when using an implicit method at each step a set of nonlinear algebraic equations

$$\bar{\Phi}(\bar{x}) = \bar{x} - h\beta\bar{f}(\bar{x}) + \bar{d} = 0 \qquad (3.6)$$

where β is some constant characteristic of the method, must be solved by some Newton technique, i.e.

$$\bar{x}_{n+1} = \bar{x}_n - (I - h\beta\partial\bar{f}/\partial\bar{x})^{-1}_{\bar{x}_n}\bar{\Phi}(\bar{x}_n) \qquad (3.7)$$

It can be shown that the Jacobian can be generated from the formula

$$\partial\dot{\bar{c}}_1/\partial\bar{c}_1 = S_1 K\;\mathrm{dg}\;\bar{c}^{-P^T}\;P^T\;(\mathrm{dg}\;\bar{c})^{-1}\begin{pmatrix} I \\ L \end{pmatrix} \qquad (3.8)$$

as long as $c_i \neq 0$, which is usually the case as soon as the first step has been taken, see Edsberg (1975B). We see that the Jacobian is easily updated; c^{-P^T} is calculated when the right hand side is computed and $\bar{c}$ is the approximate values of the components themselves, all other matrices are constant.

3.2. Instability outside the reaction simplex

In the previous chapter it was stated that (2.1) under certain circumstances (which in fact are fulfilled for most examples in practise) has a unique equilibrium point in Ω_0. When (2.1) is a linear system this point will eventually be reached from any starting point in L_0, i.e. even points where some c_{i0} are negative. Hence the equilibrium point is asymptotically stable in the whole of L_0 not only Ω_0.

This, however, is not true for nonlinear systems of type (2.1). Horn and Jackson (1972A) have shown that for a CB system the equilibrium point is globally asymptotically stable with respect to Ω_0, but not with respect to L_0. This is visualized with the simple irreversible reaction

$$2A \rightarrow B \tag{3.9}$$

the dynamics of which is described by

$$\dot{x} = -2kx^2 \; ; \; x(0) = x_0 \tag{3.10}$$

which has the solution

$$x = x_0/(1 + 2x_0 \, kt) \tag{3.11}$$

From this simple example we see that

1. The equilibrium point ($x_\infty = 0$) is stable only for positive perturbations of x.

2. Once x becomes negative for some reason the solution becomes unstable.

This 'semistability' character has turned out to be a problem for certain testexamples. For chemical problems the stiff variables often are small and as soon as they are negative due to some perturbation (e.g. rounding errors) the solution leaves Ω_0 and becomes unstable.

Example 2. The following problem occurred in practise in a model describing oxidation of propane, Björnbom (1975). The reaction system

$$\begin{aligned} A &\rightarrow B \\ A + B &\rightarrow C + B \\ C &\rightarrow B \\ B + D &\rightarrow E \\ 2B &\rightarrow 2E \end{aligned}$$

and the corresponding ODE's are after application of mass action law

$$\dot{x}_1 = -k_1x_1 - k_2x_1x_2$$

$$\dot{x}_2 = k_1x_1 + k_3x_3 - k_4x_2x_4 - 2k_5x_2^2$$

$$\dot{x}_3 = k_2x_1x_2 - k_3x_3$$

$$\dot{x}_4 = -k_4x_2x_4$$

$$\dot{x}_5 = k_4x_2x_4 + 2k_5x_2^2$$

In the special application we had the following values of the rate constants: $k_1 = 10^{-4}$, $k_2 = 2.9\cdot10^4$, $k_3 = 5\cdot10^3$, $k_4 = 10^4$ and $k_5 = 6.7\cdot10^{10}$. The initial values were $x_{10} = 0.6$, $x_{20} = 0$, $x_{30} = 0$, $x_{40} = 0.4$, $x_{50} = 0$. The general behaviour is that x_1 and x_4 slowly decrease monotonically, x_5 increases while x_2 and x_3 both reach a maximum at $t = 5.6\cdot10^{-3}$, with values 10^{-7} and $3.6\cdot10^{-7}$ respectively.

Before we continue to present the results of different runs of example 2 it is convenient to say something about the scaling problem in general for stiff systems. The following types of scaling are possible for an implicit method:

1. Scaling of the variables in the ODE's, i.e. transformation of the ODE's after application of (2.8).

2. Use of a scaled norm in the computation of the local error estimate.

3. Use of a scaled norm to measure the accuracy of the iterates obtained from the Newton-Raphson procedure.

In Lindberg's program it is possible to give scaling factors for the local error and the Newton iterations. When the problem was run without any scaling, i.e. all scaling factors were set to 1 the program immediately slipped off the solution with positive values to an unstable solution. When the local error and the errors in the Newton iterates were scaled by the factors $1, 10^7, 10^7, 1, 1$ the correct trajectory was followed but the stepsize increased very slowly. However, when the ODE's were scaled by factors so that the maximum value of each component were of the same order (= 1) the program behaved 'normally'. The same type of behaviour turned out to be present when a version of Gear's package was used. In both cases the problems were run on an IBM 360/75 in double precision. It is still an open question which type of scaling is the best in general, but numerical experiments on a lot of other test examples have shown that scaling of type 1 seems to be the safest at present, which, however, demands a priori knowledge of the magnitudes of the components.

3.3. Some results concerning the Newton-iterations

One of the heavy computational parts when solving stiff problems with implicit methods is the solution of the set of non-linear algebraic equations (3.6) in each step. To increase the efficiency in most program packages one uses the same LU-decomposed matrix

$$(I - h\beta\partial\bar{f}/\partial\bar{x})^{-1} \tag{3.12}$$

for several steps before it is updated at steplength changes or if the rate of convergence is not satisfactory. The reason why Newton iterations are used when solving (3.6) is that a direct iteration scheme

$$\bar{x}_{n+1} = h\beta\bar{f}(\bar{x}_n) - \bar{d} \tag{3.13}$$

will not converge. Only in the initial phase of the integration, where h is so small that

$$||h\beta\partial\bar{f}/\partial\bar{x}|| < 1 \tag{3.14}$$

such a method converges. However, if it is possible to make a partitioning of the set of ODE's into fast and slow components, i.e. the partitioning (3.1) we can use a mixed iterative scheme of the type

$$\begin{aligned} \bar{c}_{1n+1} &= \bar{c}_{1n} - (I - h\beta\partial\bar{f}_1/\partial\bar{c}_1)^{-1}\bar{\Phi}_1(\bar{c}_{1n},\bar{c}_{2n}) \\ \bar{c}_{2n+1} &= h\bar{f}_2(\bar{c}_{1n+1},\bar{c}_{2n}) + \bar{d}_2 \end{aligned} \tag{3.15}$$

thus the stiff components are solved by Newton-iterations while the non-stiff are solved with direct iterations. The method was suggested by Dahlquist (1974), where also some discussion of the convergence properties are discussed. Recently Robertson (1975) has suggested a similar technique.

A good behaviour of such a method when implemented in a method would be that in the initial phase all equations are solved with the direct technique, while in the asymptotic phase, where large steps can be taken, Newton-iterations are used on the whole system. In the intermediate phase the components are investigated with respect to stiffness and the partitioning (3.1) is modified successively. At present this automatic partitioning has not been implemented but in the program package by Edsberg

(1975A) this partitioning has recently been implemented in such a way that the user can give it interactively. This partitioning is then kept fixed during the whole range of integration.

We report here the experiences of this method on two test-examples, example 2 given in chapter 3.2 and the famous example by Robertson

$$\begin{aligned} \dot{x}_1 &= -k_1x_1 + k_2x_2x_3 & x_{10} &= 1 \\ \dot{x}_2 &= k_1x_1 - k_2x_2x_3 - k_3x_2^2 & x_{20} &= 0 \\ \dot{x}_3 &= k_3x_2^2 & x_{30} &= 0 \end{aligned} \qquad (3.16)$$

In example 2 x_2 and x_3 are the stiff components and in Robertson's example x_2 is stiff ($k_1 = 0.04$, $k_2 = 10^4$, $k_3 = 3 \cdot 10^7$). Some of the results obtained are given below. In the table A means accuracy, B the number of LU-decompositions, C the maximum number of iterations needed in a step, D the number of steps to t_e = end of integration interval and E the partitioning used.

Example	A	B	C	D	E
Robertson	10^{-5}	13	4	20 $t_e = 0.05$	x_2 stiff
	10^{-5}	13	2	20 $t_e = 0.05$	whole system
Example 2	10^{-4}	14	2	14 $t_e = 0.01$	x_2, x_3 stiff
	10^{-4}	16	2	14 $t_e = 0.01$	whole system

We see from this that this mixed iterative scheme may be quite efficient compared to using Newton-iterations on the whole system. In example 2 this mixed method needed to invert only a 2×2 matrix instead of a 5×5 matrix and yet the number of iterations were not increased. Test runs with this method in the asymptotic region, however, showed that the condition (3.14) was violated by the stepsize regulation; too large steps were taken with respect to this condition so it would have been advantageous to switch to Newton-iterations on the whole system somewhere in the intermediate phase. Work on this is made at present.

4. CONCLUSIONS

The problem class of mass action kinetics contains examples of problems which can be very stiff and also exhibit other numerical difficulties which must be taken into consideration in the implementation of a method. But there are also certain short-cuts to be taken. Often the poser of a chemical kinetics problem has some a priori knowledge about which components are fast and which are slow. In such a case a special type of iterative scheme seems to be worth trying since the amount of work to be excepted is much smaller than if a general purpose program is used.

5. ACKNOWLEDGEMENTS

This work was supported by the Swedish Institute of Applied Mathematics. The author wishes to thank Prof. G. Dahlquist for many stimulating discussions.

REFERENCES

Aiken R.C., Lapidus L. "An effective numerical integration method for typical stiff systems", A.I.CH.E. 20 no 2,368 (1974).

Aris R. "Introduction to the analysis of chemical reactors", Prentice Hall (1965).

Aris R. "Mathematical aspects of chemical reaction", Ind. Engn. Chem. Fund: 61, 17 (1969).

Banford C.H., Tipper C.F.H. "Comprehensive chemical kinetics" Banford, Tipper eds, vol. 2 "theory of kinetics" Elsevier (1969).

Björnbom P. "The kinetic behaviour of the liquid phase oxidation of propane in chloro-bensene solution", Dissertation, Dep of Chemical Technology, The Royal Institute of Technology, Stockholm (1975).

Clark R.L., Groner G.F. "A CSMP/360 precompiler for kinetic chemical equations", Simulation, 127, oct (1972).

Curtis A.R., Chance E.M. "Chek and Chekmat; two chemical reaction kinetics programs", A.E.R.E. Rep 7345, Harwell England (1974).

Dahlquist G. "Problems related to the numerical treatment of stiff differential systems", in "ACM Proc. of Int. Comp. Symp.", North Holland (1974).

Edsberg L. "Integration package for chemical kinetics" in "Stiff differential systems (Willoughby R ed) Plenum Press (1974).

Edsberg L. "KEMPEX II, A program package for interactive simulation of some chemical reactors", Rep TRITA-NA-7504, Dep. of Inf. Proc. Computer Science, The Royal Inst of Techn Stockholm Sweden (1975A).

Edsberg L. "Some mathematical properties of mass action kinetics" Rep TRITA-NA-7505, Dep. of Inf. Proc. Computer Science, The Royal Inst of Techn Stockholm Sweden (1975B).

Enright W., Hull T., Lindberg B, "Comparing numerical methods for stiff ODE's" BIT 15 10-48 (1975).

Feinberg M. "On chemical kinetics of a certain class", Arch. Rat. Mech. Anal. 46,1 (1972).

Feinberg M., Horn F.J.M. "Dynamics of open chemical systems and the algebraic structure of the underlying reaction network", Chem. Engng. Sci. 29, 775 (1974).

Garfinkel D. "A machine-independent language for simulation of complex chemical and biochemical systems", Comp. and Biom. Res. 2, 31 (1968).

Gavalas G.R. "Nonlinear differential equations of chemically reacting systems", Springer (1968).

Grover S.S. "Application of computers in chemical reaction systems", Cat. Rev. 1, 153 (1967).

Himmelblau D.M., Bishoff K. "Process analysis and simulation: deterministic systems", Wiley (1968).

Horn F. "Necessary and sifficient conditions for complex balancing in chemical kinetics", Arch. Rat. Mech. Anal. 49, 172 (1972B).

Horn F., Jackson R. "General mass action kinetics", Arch. Rat. Mech. Anal. 47, 81 (1972A)

Laurendeau N., Sawyer R.F. "General reaction rate problems: combined integration and steady state analysis", Rep. no TS-70-14. Thermal Systems division, Dep. of Mech. Engn., University of California (1970).

Lindberg B. "Package based on the implicit midpoint method" in Stiff differential Systems" (Willoughby R. ed) Plenum Press (1974).

Oster G.F., Perelson A.S. "Chemical reaction dynamics" Arch. Rat. Mech. Anal. 55, 230 (1974).

Robertson H.H. "Numerical integration of systems of stiff ordinary differential equations with special structure", Numerical Analysis rep no 8 march 1975, Dep. of Mathematics, Univ. of Manchester England (1975).

Snow R.H. "A chemical kinetics computer program for homogeneous and free radical systems of reactions", J. of Phys. Chem: 70, 9, 2780 (1966).

Wei J. "An axiomatic treatment of chemical reaction systems", J. Chem. Phys. 36, 1578 (1962).

Wei J., Prater C.D. "The structure and analysis of complex reaction systems", Adv. Catalysis 13, 203 (1962).

A Systematized Collection of Codes for Solving Two-Point Boundary-Value Problems

M. R. Scott and H. A. Watts

1. Introduction

The development of quality software is both time consuming and expensive. This is particularly true in the case of boundary-value problems since it is difficult to imagine one algorithm ever being able to solve all problems. Some of the types of problems encountered are linear and nonlinear, two-point and multipoint, determined and overdetermined boundary conditions, small and large systems, singular coefficients and singular intervals, slowly and rapidly varying components, and eigenvalue problems. It indeed would be remarkable if one algorithm could be developed which could handle all of these possibilities both accurately and efficiently.

There appear to be very few readily available codes for solving boundary-value problems in ordinary differential equations. Lentini and Pereyra [1] have developed a very general finite difference code which uses deferred corrections and an automatic mesh generation procedure. Indications are that future generations of this code will lead to a powerful tool for solving

This work was supported by the U.S. Energy Research and Development Administration.

very general boundary-value problems. England, Nichols, and Reid [2] and Bulirsch, Stoer and Deuflhard [3] have developed multiple shooting codes which use efficient variable step integrators. Both codes allow either a pre-assigned set of shooting points to be used or the shooting points to be selected automatically. However, it is our understanding that neither of the codes dynamically select these shooting points. Rather, a sort of preprocessing pass at solving the problem is made, at which time some criteria are assessed about where the shooting points should be placed. Then, these points do not change for the remainder of the process which attempts to solve the resulting nonlinear equations by some variation of Newton's method. Scott and Watts [4] describe a code for solving linear problems using the principle of superposition coupled with an orthonormalization procedure. The code uses effective initial-value methods and automatically determines when orthonormalization should take place. The orthonormalization points are introduced for the purpose of maintaining numerical linear independence of solutions in the superposition process and, in some ways, roughly correspond to the selection of shooting points.

In view of these comments we have embarked upon a project of writing a systematized collection of codes for the numerical solution of boundary-value problems which will routinely handle many of the above classes of problems and yet be convenient and easy to use. In this paper, we shall describe only those codes which are presently available or nearing completion. These consist entirely of procedures which utilize initial-value methods. Among these are an orthonormalization procedure, an invariant imbedding algorithm, and several versions of shooting. Although the orthonormalization and invariant imbedding procedures are designed to solve linear problems they can also be used to solve certain types of nonlinear problems when combined with quasilinearization. Due to the excellent state of the art development of initial-value codes and the small amount of storage required, we consider such techniques to be very attractive.

In Section 2 we discuss the methods presently being considered for solving the two-point boundary-value problem. The initial-value codes being used are discussed in Section 3 and, in Section 4, we describe the boundary-value codes which we have developed. Lastly, the boundary-value codes' performances on several example problems are presented in Section 5.

2. Methods

The general problem considered is of the form

$$y'(x) = f(x,y(x)) \tag{1}$$

subject to the two-point boundary conditions

$$Ay(a) + By(b) = \gamma \tag{2}$$

where y and γ are vectors with n components, and A and B are matrices of order n such that the rank of the n x 2n matrix $[A, B]$ is n. If the problem is linear, Eq. (1) is normally rewritten in the form

$$y'(x) = F(x)y(x) + g(x) \ . \tag{3}$$

Only methods which convert the boundary-value problem into an initial-value problem are examined. There are several reasons for this choice. There are a number of sophisticated codes which have been developed recently for solving initial-value problems; these will be discussed in the next section. The authors have not written a finite difference code since Lentini and Pereyra are developing an excellent code. Collocation and Galerkin methods were not considered at this time since there are no efficient ways of producing a variable net appropriate for these procedures. Much work must be done in this area before these methods are as attractive as the initial-value methods.

Once it was decided to use procedures which use initial-value methods, this category had to be narrowed even further. It was determined to use a technique based upon superposition combined with an orthonormalization procedure [4, 5, 6] (Appendix B of [4] contains an English translation of [5]), an invariant imbedding algorithm [7, 8], and a shooting algorithm. The first two methods are specifically designed to solve linear problems while the shooting algorithm will handle both linear and nonlinear problems. In this report, shooting is defined as the procedure where missing initial conditions are guessed at, the resulting initial-value problem is solved, and some iterative scheme is used to solve the transcendental equations in the unknown parameters -- thus generating a sequence of initial-value problems. Of course, the use of Newton's method in a shooting technique applied to a linear problem is theoretically a two pass integration process, the first for obtaining the correct missing conditions and the second for producing a solution to the boundary-value problem. However, in practice a few additional iterations are often required. Aside from the use of numerical differences in derivative free Newton-like schemes, this is due in part to checking to see if convergence has been satisfactorily achieved and to the sensitivity of the solution with respect to small changes in initial conditions. Attention is actually restricted to a quasi-Neton scheme similar to that described in [9]. In the invariant imbedding and orthonormalization algorithms, only one integration pass is needed although additional initial-value problems are being solved. These solutions are then combined to form the solution of the boundary-value problem. We wish to separate from the general context of shooting those procedures not utilizing an iterative scheme. It is felt that the connotation of shooting is not descriptive enough and the commonality of the algorithms should instead be considered to lie in the use of initial-value methods.

The method of superposition is perhaps the simplest of all techniques for solving linear boundary-value problems. The algorithm proceeds by producing linearly independent solutions of the homogeneous equation and a particular solution of the inhomogeneous equation. The solution of the original problem is then produced by forming an appropriate linear combination of the solutions of the homogeneous and inhomogeneous equations which satisfies the given boundary conditions. This procedure, in principle, is quite simple. However, there are several well-known numerical difficulties associated with superposition. First, due to the finite word-length of the computer, it may be difficult to produce linearly independent solutions of the homogeneous equation. Secondly, even if linear independence is maintained, a loss of significance can occur in the formation of the linear combination. In order to overcome these difficulties and yet retain the simplicity, superposition is combined with an orthonormalization procedure. The basic idea is that each time an impending loss of independence is detected, the solution vectors are orthonormalized and the integration continued with the new set of initial conditions. The "orthonormalized" solutions are then pieced together over the various subintervals by means of a simple recursion process. An important feature of the algorithm is the automatic determination of when to orthonormalize.

There are several invariant imbedding algorithms. A recent survey [8] indicates that the algorithm developed by Scott has the most potential for a general purpose code. The algorithm proceeds by relating a new set of dependent variables with $y(x)$ via a Riccati-like transformation. These new variables satisfy a set of initial-valued nonlinear differential equations which are generally more stable than the original set. After the new differential equations are integrated over the desired interval, the solution of the original boundary-value problem can be obtained by solving a set of linear algebraic equations.

Both the orthonormalization and the invariant imbedding procedures are based upon solving linear equations. In order to apply these two algorithms to nonlinear problems, the method of quasilinearization [4, 7, 10] is applied. This way of linearization is actually an application of Newton's method to the nonlinear differential operator. This results in a sequence of linear boundary-value problems which may be solved using either orthonormalization or invariant imbedding.

An often overlooked difficulty of such iterative procedures is the necessity of providing a solution of the previous iteration everywhere it is required. Obviously, if the integration method were restricted to use the same mesh (e.g., a uniform selection) from one iteration to the next, this difficulty would not exist. But since efficient variable step integrators are being used, extra care must be given to this problem. The obvious approach is to use interpolation. However, the importance of the interpolation process in producing very smooth results (sufficiently high order continuous derivatives) everywhere in the interval should be stressed. Otherwise, the efficiency of the integration scheme will be seriously impaired. Since solution values and derivatives are readily available, Hermite interpolation seems a natural choice. Besides the caution which should be taken to ensure smoothness across subintervals where different interpolating polynomials are defined, there is also the problem of knowing where to place the interpolation joints. This can have important implications on the accuracy which can be achieved for the solution of the boundary-value problem.

Another very interesting way of providing this information about the solution on the non-mesh points is to integrate the original nonlinear equation. This is the procedure the authors prefer. On difficult problems, it may not be possible to integrate the original equation over the total interval. In such cases, the integration must be restarted at appropriate points such as designated output points, orthonormalization points, or points chosen by some other means. The initial conditions for the nonlinear problem are obtained from the solution of the previous iteration. This process can be automated easily, but it turns out that additonal controls must be placed on the intermediate integrations for the procedure to converge. This is reasonable since practical controls are necessary in attempts to force Newton's scheme to converge from bad starting guesses. It would not serve any purpose to discuss these ideas further since they can be improved upon.

The shooting codes developed by the authors use an algorithm which, although quite simple, is usually not explicitly stated in the literature. Consider an associated initial-value problem

$$u'(x) = f(x, u(x))$$

$$u(a) = s$$

where s is an arbitrary set of initial conditions. The idea of shooting is to find a vector s such that

$$\Phi(s) = As + Bu(b,s) - \gamma = 0.$$

Since in general this is a transcendental equation in s, the standard method is to use Newton's scheme to adjust s until convergence is obtained [11]. This results in

$$s_{i+1} = s_i - Q^{-1}(s_i)\Phi(s_i)$$

where $Q(s) = A + BW(b,s)$ and $W(b,s)$ is the Jacobian matrix, i.e., $W(x,s) = \frac{\partial u(x,s)}{\partial s}$ and is obtained by solving the associated variational equations. This necessarily results in obtaining analytical partial derivatives, which can be tedious even for rather simple-looking problems. We have chosen to use a quasi-Newton scheme which uses numerical approximations for the $Q^{-1}(s_i)$. Although the convergence of the modified Newton's scheme is less than quadratic, it is superlinear and the convenience of not having to obtain the analytical Jacobian overrides the possibility of a slower convergence rate. However, it is also likely that the modified scheme will have a faster execution time because of the possibility of fewer integrations being performed. The reason is that the quasi-Newton method uses Broyden's rank one update for approximating Q^{-1} [9, 12]. This often results in an acceptable Q^{-1} being formed from only a single evaluation of Φ. In this case, one integration of the original differential equation system results in a Newton-type iteration. On the other hand, the application of the analytical Newton approach requires the solution of n additional systems of linear differential equations for each Newton iteration.

On difficult problems simple shooting may prove to be inadequate. For example, we may be unable to perform the integration over the full interval. A remedy that will often work is the idea of multiple or parallel shooting in which breakpoints are introduced at appropriate places in the original interval [11]. The advantages of multiple shooting over simple shooting are lower error growth rates due to the shortened subintervals of integration and an enlarged domain of attraction for convergence of the Newton scheme. In fact, there is an exponential change in the theoretical radius of convergence [13]. We have implemented multiple shooting only from the viewpoint of using preassigned breakpoints because we do not have the automatic determination of these points operational.

The root solving technique presently employed for the shooting algorithms does not take advantage of any special structure of A and B. In particular, in the context of multiple shooting there is an approximately block bidiagonal type structure arising in both A and B. However, the failure of this technique to take special account of such features does not create a gross inefficiency with its use. The reason for this lies with a quasi-Newton scheme being used for the root solving. An approximation to the inverse matrix Q^{-1} is updated each iteration step, and this is accomplished by a single integration of the differential equation system. Since most of the computational effort is generally associated with the integration problem in such circumstances, the application of this technique is simply not as inefficient as it might otherwise seem. However, if the quasi-Newton root solver is having difficulty with global convergence and finds it necessary to completely reinitialize the approximation to Q^{-1} frequently, this algorithm does become inefficient whenever a sparse structure is present in A and B. Reinitialization differs from the updating procedure in that n evaluations of Φ are required for initialization of Q^{-1} since difference approximations are used to compute the Jacobian matrix. This corresponds to n separate integrations of the differential equation system. If band or block structure is present in the definition of Φ, this initialization can be accomplished in fewer than n evaluations. We are presently structuring the algorithm so as to take full advantage of the sparseness pattern encountered with multiple shooting. Another attractive approach in these circumstances would be the direct application of updating and reinitializing of the Q matrix, thereby preserving the sparse structure. Unfortunately, this approach now requires the frequent factorization of Q. Because of this, it would seem that a rather large system is necessary before this algorithm can become more effective than the first. These possibilities will be examined in more detail at a later date.

Another rather interesting phenomenon which we have observed on numerous occasions is the following. The iterative scheme of applying Newton's method (quasilinearization) to the differential operator appears to have a larger domain of convergence than does the iterative scheme of applying Newton's method directly to the function defining the unknown parameters (shooting). The latter approach solves nonlinear differential equations to approximate the partial derivatives needed in the iteration function. The first approach solves a sequence of linear differential equations, with the iteration now basically taking place in function space. It would be interesting if results or conditions could be established under which one approach would have a decided advantage.

3. Integration Methods

Since the procedures we have discussed for solving boundary-value problems lead to initial-value problems, we shall take advantage of the current state of the art and use some sophisticated initial-value integration codes. This is in contrast to many applications discussed in the literature where only fixed stepsize integration schemes have been utilized. Such schemes tend to be inefficient or inaccurate, for if the stepsize is too small unnecessary computation is being performed, while too large a stepsize means that the desired accuracy probably is not being achieved. Because the systems of differential equations to be solved for any given problem may involve rapidly varying functions, it is necessary to incorporate some type of variable step control in the integration procedure. Furthermore, if the equations are quite complicated and costly to evaluate we need a method which minimizes the number of function evaluations. Hence, very effective integration methods coupled with accurate determination of the stepsize to control the error in the integration process are essential for the overall procedure to progress efficiently and reliably to a solution.

For the initial-value codes we use, the stepsize is adjusted so as to keep an estimate of the local error below the user specified tolerance. Often there is confusion on the part of users as to what error is being measured or controlled and what is its relation to the accuracies requested by users and those actually achieved. This can be further complicated here by the fact that solutions to several initial-value problems may be recombined to form the solution of the boundary-value problem. The effect on the global error of an initial-value solution by keeping the local error small depends on the differential equation itself, but unless the equation is mathematically unstable and the interval of integration is too long, controlling the local error roughly controls the global error.

We shall briefly discuss the codes which are used. Except for the stiff solver, these are entries from the systematized collection of codes for solving ordinary differential equations, DEPAC, now being written at Sandia Laboratories -- some of which are presently available from the Argonne Code Center.

The first integrator we mention is based on the Runge-Kutta-Fehlberg pair of fourth and fifth order formulas which use six derivative function evaluations per basic step [14]. We shall refer to the code here as RKF although it is essentially the same as the RKF45 version in DEPAC which was used in the tests [15] and described in [16]. This is a very efficient fifth order method (by virtue of applying local extrapolation to a fourth order process).

The second integrator which we have available uses the same basic Runge-Kutta process as in RKF except that the integration scheme now carries two parallel solutions, one using half the stepsize of the other, to estimate the global error. This code is called GERK and applies global extrapolation for estimating the global error in the more accurate solution which is reported to the user [17]. The global error estimates supplied by the GERK routine are used in the reorthonormalization and multiple shooting procedures for determining the interval break points where new initial-value problems are defined.

A detailed comparison between RKF and GERK can be found in [17]. One situation in which GERK will not perform in the expected manner of RKF is worth mentioning; this is with respect to discontinuous derivative functions. Because the stepsize and error control is based on the approximation over the larger step, the reported solution could staddle a slightly shifted point of discontinuity, resulting in a less accurate value than expected. This phenomenon cannot occur if the point of discontinuity is determined solely by the independent variable, but the possibility exists if the dependent variables define the discontinuity. Therefore, when this situation is present we urge the usage of one of the other integrators.

The third integrator consists of the Adams package DE/STEP, INTRP [18]. This is a variable order method which uses a modified divided difference formulation of the Adams PECE formulas and local extrapolation. The formulas go up through order 12. A detailed discussion of this method and code is found in [18].

When solving linear problems, the Adams code can be utilized in a way that only one-half the usual number of matrix and vector derivative evaluations need be performed. If we are solving a problem where the matrix and/or vector components are exceedingly expensive to compute, a considerable savings could result from the use of an Adams code in lieu of a Runge-Kutta code. In general, if we are asking for high accuracies, an Adams code will be capable of achieving them more efficiently (in terms of frequency of derivative function evaluations). However, overhead can be significant on simple equations. A rather complete comparison between DE and RKF (and others) is found in [15].

The fourth integrator is a code for solving stiff differential equations. Although some authors call a problem stiff if the eigenvalues of the associated Jacobian matrix are widely separated, allowing positive values to be included, we prefer a more restrictive definition which requires the eigenvalues to lie in the left half-plane. Specifically, the system of linear differential equations (3) is said to be stiff if all the eigenvalues μ_i of the matrix F have negative real parts and max $|\mathrm{Re}\mu_i| \gg \min|\mathrm{Re}\mu_i|$. From a computational point of view, there is an important distinction between these two definitions. When the eigenvalues have positive real parts, the differential equations are mathematically unstable and the integrator restricts its stepsize in order to maintain accuracy in the solution. However, if all the eigenvalues have negative real parts, the differential equations are stable and, typically, the solution behavior would allow large steps to be taken. Unfortunately, with most integration methods such as the Runge-Kutta and Adams methods, the stepsize must be severely limited or else numerical instability occurs. This stability restriction on the stepsize applies over the entire integration interval -- a rather severe penalty if the step length is excessively small relative to the interval length. This is the problem of stiffness. To solve stiff problems efficiently, one must resort to special numerical methods such as those available in Gear's code [19]. We shall refer to our Sandia implementation of the stiff methods as STIFF.

The user of the boundary-value codes requests the particular integration code to be used and the error tolerances to be attempted. These tolerances, the relative and absolute error parameters RE and AE, respectivley, are then used by the particular integrator for stepsize adjustment in keeping the estimated local error per step below the acceptable tolerance RE · (solution) + AE.

4. Codes

As the title of this report indicates, we have developed a collection of computer codes for solving two-point boundary-value problems. We refer to them as being "systematized" in the sense that the arguments of call lists of codes in the collection will be as near to being identical as possible in both name and usage. Furthermore, we have aimed at a consistent philosophy of the tasks undertaken and the level of protection afforded among all the codes. We have also attempted to make the codes as portable as possible. The basic external appearance of the code design emphasizes user convenience while maintaining sufficient flexibility to handle a variety of problems.

As previously noted, our collection of boundary-value codes is not complete. To date, we have developed codes based only on utilizing initial-value procedures. The justification for having a collection of codes (using different methods) lies in our belief that no one technique will be best for all problems. A particular method may be incapable of producing an efficient solution (if at all) on certain types of problems. The justifications for having several shooting codes are efficiency and ease of use, the codes being tailored to specific boundary conditions. Such tailoring should be particularly appealing to users who wish to solve their problem with a minimal effort on problem stepup.

While we have not yet undertaken exhaustive comparisons between the codes included here, we will attempt to provide enough information about each code to enable a prospective user to make a choice. For a collection of codes that solve the same problem to be generally useful, some directives for picking a code are necessary. On the other hand, it is sometimes just comforting to have independent techniques readily available when solving difficult problems. This gives the user a framework within which he can experiment. Being able to easily make such a comparison check is a basic aim of systematizing the collection.

We shall now list the codes which we have developed and comment briefly upon their use and special features. The SUPORT code (using superposition and orthonormalization) for solving linear problems has received most of our attention. The method and important implementation features are discussed in detail in [4]. The present version of this code solves only problems with separated boundary conditions,

$$Ay(a) = \alpha, \quad By(b) = \beta,$$

where now A is an (n-k) x n matrix of rank n - k and B is a k x n matrix of rank k. The user must supply the information which defines the problem: boundary condition matrices A, B and vectors α, β and a subroutine for evaluating the differential equation. In SUPORT this is split apart, with the user writing two subroutines - one to produce the vector Fy and the other to produce the vector g. The splitting apart for evaluating the derivatives is necessary for this algorithm since we are computing solutions to both the homogeneous and inhomogeneous equations. Because all equations are being solved simultaneously as one larger system a considerable savings can be affected when F(x) is expensive to evaluate. This can be achieved by the relatively simple programming device of using COMMON blocks to save the complicated parts of F(x) and reusing them as long as the independent variable x remains unchanged.

The user must also provide a requested error tolerance in the form of relative and absolute error parameters which are used by the initial-value integrators. Also, the user chooses the direction of integration since the code proceeds from a to b. This can be quite important in achieving an efficient solution. Lacking additional information about the growth characteristics of the equation, one should normally integrate toward the point having fewest specified boundary conditions. This results in the smaller number of initial-value problems to be solved. The user is not burdened with supplying the initial conditions for the independent vector set. Rather, SUPORT automatically chooses values which satisfy the initial boundary conditions $Ay(a) = \alpha$.

All the codes, except the current version of INVIMB, are designed to handle an arbitrary number of differential equations although there is naturally an upper limit due to storage availability. However, the user is not burdened with providing all the intermediate arrays which are necessary in such sophisticated procedures. Rather he provides only two such arrays, real and integer, to act as the working areas needed. The code splits apart these arrays and distributes the work space as dictated.

The SUPORT code has proven to be very reliable and reasonably efficient in solving moderately sized systems of linear differential equations. The code is primarily aimed at solving such problems which may be classified as mathematically unstable or which exhibit a strong sensitivity to initial conditions via the integration process. The success of SUPORT as a useful piece of software depends on effective integration and the automatic choice of interval subdivision (orthonormalization points). The code has two such schemes available which we have found to be both reliable and efficient in the sense of not introducing an abundance of unnecessary breakpoints. In SUPORT, certain information must be stored at these points.

The INVIMB code (using invariant imbedding techniques) is currently restricted to solving only second order linear problems although work is underway for a general linear system code. Here the imbedding process requires the matrix $F(x)$ to be available. In fact, for a general system the Riccati equations depend on certain factorizations of $F(x)$. For the second order code, function subprograms must be supplied by the user to evaluate the elements of $F(x)$. If the Riccati equation has a singularity on the interval, it is necessary to switch to another Riccati transformation (e.g., the inverse) before continuing the integration. This is handled automatically in the code as is the technique of successive restarts designed to overcome possible loss of significance in the recombination phase of the algorithm.

At the time of this writing, four basic shooting codes are available -- SHOOT1, SHOOT2, SHOOT3, and SHOOT4. As previously mentioned, these were designed specifically for the convenience of the user on problems with certain boundary conditions. In particular, SHOOT1 was written for conditions of the form

$$y(a) = \alpha, \; y(b) = \beta;$$

SHOOT2 for conditions of the form

$$y(a) = \alpha, \ By(b) = \beta;$$

SHOOT3 for conditions of the form

$$Ay(a) = \alpha, \ By(b) = \beta;$$

and SHOOT4 for conditions of the form

$$Ay(a) + By(b) = \gamma.$$

In the above we consider the direction of integration to proceed from a to b. All of these codes are operational currently with the simple shooting procedure. Multiple shooting aspects (with automatic determination of breakpoints) are being investigated but are not operational at this time. Obviously, SHOOT4 can be used as a multiple shooting code provided the breakpoints are assigned in advance. This requires the user to construct the appropriate new system of differential equations defined on a common interval, say [0, 1].

As indicated in the discussion on shooting, SHOOT4 does not take advantage of any special structure of A and B, such as block diagonal features. Nevertheless, in many such applications SHOOT4 will not be terribly inefficient. In particular, when used as a multiple shooting code SHOOT4 will perform quite nicely provided reasonable initial guesses can be given so that the quasi-Newton scheme [12] can avoid reinitializations. We are presently working on another shooting code which is specifically designed to handle multiple shooting.

The nonlinear extension of the basic SUPORT code is called SUPOR Q (<u>sup</u>erposition and <u>orth</u>onormalization coupled with <u>q</u>uasilinearization). Currently this code is in a very preliminary stage of development. Although there is a considerable amount of fine tuning yet to be achieved, we have obtained quite satisfactory results from this version. We have already commented on the important aspect of providing a solution of the previous iteration everywhere it is needed. SUPOR Q is written with this information being basically provided by integrating the original nonlinear problem. However, we do show some comparisons with an earlier version which used interpolation. Most of our emphasis in the collection of codes has been directed at the SUPORT code for linear problems and now at SUPOR Q for nonlinear problems. The major distinction between use of the two codes comes in defining the differential equation subroutines. For the nonlinear problem, the user is confronted with computing the Jacobian $\frac{\partial f}{\partial y}$, a sometimes tedious and error prone task at best.

In fact, the present code requires subroutines for evaluating $\frac{\partial f}{\partial y}$ (u) y and f(u) - $\frac{\partial f}{\partial y}$ (u) u, where u represents the solution for the previous iteration. This is not particularly convenient from the user's standpoint but appears to be more efficient than merely supplying the Jacobian matrix definition. The latter approach would involve matrix-vector multiplications, frequently with many zero elements. More experience and feedback on the merits of these two approaches will be needed before a final design can be settled upon. We also anticipate building in some sort of continuation procedure, a very important technique necessary for solving some problems.

On each iteration, the SUPOR Q code computes k + 1 independent solutions: a particular solution to the inhomogeneous problem and k solutions which are sufficient to form the basis of the homogeneous problem space [4]. Since all equations are integrated simultaneously, a considerable savings can be achieved whenever $F(x) = \frac{\partial f}{\partial y} [u(x)]$ is expensive to evaluate; that is, only one evaluation of F will suffice for the k + 1 solutions. The importance of this observation is seen when function evaluations are used as a measure of the efficiency of codes. In fact, when derivative evaluations become the primary cost the user should program the subroutine where F is being computed in the manner suggested in the discussion on SUPORT. In such circumstances, the effective cost becomes: (total number of derivative subroutine calls)/(k + 1). (This should be kept in mind when examining the tables of numerical examples. The numbers of derivative evaluations shown there should be divided by k + 1 to obtain an effective cost.)

In general, when simple shooting works <u>well</u> it is often the most efficient procedure in both storage and execution time. Granted that we have no reason to suspect a priori that the problem will be difficult to integrate or is peculiarly sensitive, we may wish to try a shooting code. In the case of complicated nonlinear equations, an added advantage or convenience to the user with the shooting codes presented here is not having to obtain analytic partial derivatives for use in the linearization process. This might conceivably be an important item from a user standpoint in choosing from among the various codes. Also, for nonlinear problems, simple shooting requires only a guess of some missing conditions at a single point whereas the use of quasi-linearization coupled with linear techniques requires an initial profile of all solution components at points throughout the interval. In a physical application it is quite possible that one has reasonable guesses for the initial conditons. However, it is unlikely that one knows even a rough behavior of all solution components on the interval. Obviously, the introduction of many breakpoints in multiple shooting also amounts to providing a rough initial profile of all solution components.

For second order linear equations, numerical evidence appears to indicate a preference for INVIMB whenever the solution of the problem has a lot of structure, e.g., see [8]. In fact, the invariant imbedding approach of integrating a transformed problem (nonlinear Riccati equations) seems best when the original problem is most difficult. That is, the new equations are typically quite stable mathematically and allow for very efficient integration. The possibility that the equations may become stiff must be realized and a stiff solver used in these circumstances.

The solution of nonlinear problems can be quite challenging. While one of the shooting codes will work successfully in many instances, we consider the SUPOR Q code to be more reliable in general. In particular, it has been successfully used to solve quite difficult problems rather routinely. The most serious drawback in its use is the determination of an initial solution profile over the entire interval. Frequently, it is satisfactory to simply provide linear representations which satisfy the boundary conditions. In some instances it may be necessary to utilize a continuation procedure before convergence to the desired solution is achieved.

In summary, all the codes discussed here are still undergoing modifications as we continue to strive for improvements in performance, flexibility coupled with ease of use, and in the overall design for systematization of the codes. Quality software typically evolves over a period of some time. We believe we have taken a substantial step in the direction of generating useful production type codes for solving two-point boundary-value problems. In the next section we present some results obtained by these codes on a set of sample problems.

5. Numerical Examples

In this section, we present the results of our codes on several problems which range from very simple to very difficult. We have included several comparisons among the codes, but we wish to emphasize that the purpose of the comparisons is only to give the reader insight into the relative efficiencies of the various codes and not definitive comparisons. Also, we re-emphasize that the basic intents of our code designs are user convenience and reliability as opposed to speed of computation.

All of the examples were run in single precision on a CDC-6600 at Sandia Laboratories using the FUN compiler. The working wordlength for single precision is approximately 14 decimal digits. The computer times are in seconds and were obtained using the internal clock. Due to the multiprogramming environment of the CDC-6600, we have found that a maximum variation of approximately 5 percent could occur from one execution to the next. Except when noted otherwise, all results were obtained using the RKF integrator. For some of these problems we see that the use of the Adams code DE results in many fewer derivative evaluations, in particular for tolerances 10^{-8}, 10^{-10}.

Throughout the tables of this section the following notation is employed.

TOL - The tolerance used by the integrator to control the local error. Unless otherwise stated, we have used TOL = relative error tolerance = absolute error tolerance.

TIME - The computer time in seconds required to solve the problem on a CDC-6600.

NIS - The number of successful integration steps used by the integrator. On nonlinear problems we show the number used on the final integration of the iterative scheme.

NDE - The number of derivative evaluations. For nonlinear problems this is the total for all iterations. This count does not reflect the reduced computational effort associated with derivative evaluations being performed at the same independent variable. For SUPORT and SUPOR Q this is particularly relevant. See discussions in the Integration Methods and Codes Sections. Hence, the reader is cautioned not to attach too much significance to the absolute numbers shown.

NO - The number of orthonormalizations performed by the SUPORT code.

MAX ABS (REL) ERROR in SOL - When the exact solution is known, we have recorded the maximum absolute (relative) error in the solution (first component) which was seen at the output points. We have stated the number of output points for each example and, unless otherwise stated, the output points are equally spaced throughout the interval.

NR - No results are available.

$a(\pm b) = a \cdot 10^{\pm b}$.

Example 1:

The first example was designed by Bulirsch and Stoer [3] as a problem which is troublesome for superposition but can be solved by multiple shooting techniques. It has also been discussed by Lentini and Pereyra [1] using finite differences, and by Russell [20] using a collocation procedure. The equation is

$$y''(x) - 400y(x) = 400 \cos^2 \pi x + 2\pi^2 \cos 2\pi x$$

and is subject to the boundary conditions

$$y(0) = 0 = y(1) .$$

The exact solution is given by

$$y(x) = \frac{e^{-20}}{1+e^{-20}} e^{20x} + \frac{1}{1+e^{-20}} e^{-20x} - \cos^2 \pi x \quad .$$

The problem is only moderately difficult, and a number of techniques perform nicely.

Table I lists the results of the SHOOT1 code for several tolerances. Using SHOOT1 leads to solving a scalar equation for the unknown value $y'(0)$. The effects of the exponential growth factor are clearly visible from the errors achievable, indicating the difficulty with simple shooting. All errors shown for this problem were obtained from a sample of eleven equally spaced output points. The number of iterations required was either 2 or 3.

Table II shows the results of introducing one shooting point (at $x = 0.65$). Using SHOOT4 then leads to solving four equations for the missing initial and matching conditions. While this leads to more computational effort than with SHOOT1, better accuracy is achieved. We might expect continued improvement in accuracy from the introduction of even more shooting points. However, this is not guaranteed and usually depends on the placement of the breakpoints; this emphasizes the need for the automatic determination of these points by a multiple shooting code. The last two lines in Table II reflect the use of the Adams code. This comparative behavior is fairly typical for stringent tolerances although it should be kept in mind that actual cost ratios are very much problem dependent. The number of iterations required was from 4 to 6.

TABLE I

SHOOT1

TOL	TIME	NIS	NDE	MAX ABS ERROR-SOL
10^{-4}	0.23	35	1077	6.1
10^{-6}	0.51	80	2625	2.3(-02)
10^{-8}	1.65	194	8728	2.0(-04)
10^{-10}	4.07	479	21822	1.8(-08)

TABLE II

SHOOT4

TOL	TIME	NIS	NDE	MAX ABS ERROR-SOL
10^{-4}	0.56	34	2736	5.7(-05)
10^{-6}	1.43	77	7389	1.8(-05)
10^{-8}	3.43	185	18002	3.1(-07)
10^{-10}	9.12	457	48414	4.0(-09)
- - -	- - -	- - -	- - -	- - -
10^{-8}	2.17	NR	3512	1.5(-06)
10^{-10}	3.35	NR	5334	1.4(-08)

Table III lists the results of the SUPORT code, and Table IV lists the corresponding results of INVIMB. Note how much more expensive even the simple shooting code is on this problem. The last two lines of Table III show the results obtained using the Adams integrator.

TABLE III

SUPORT

TOL	TIME	NIS	NDE	NO	MAX ABS ERROR-SOL
10^{-4}	0.13	43	550	1	5.5(-06)
10^{-6}	0.29	94	1162	1	1.2(-07)
10^{-8}	0.68	238	2900	1	1.2(-09)
10^{-10}	1.66	603	7270	1	6.5(-11)
- - - -	- - - -	- - - -	- - - -	- - - -	- - - -
10^{-8}	0.42	142	572	1	1.3(-09)
10^{-10}	0.62	206	830	1	3.7(-10)

TABLE IV

INVIMB

TOL	TIME	NIS	NDE	MAX ABS ERROR-SOL
10^{-4}	0.09	28	NR	1.1(-03)
10^{-6}	0.15	59	NR	2.3(-06)
10^{-8}	0.31	133	NR	4.5(-08)
10^{-10}	0.70	325	NR	5.1(-10)

Example 2:

The second example we wish to consider is the nonlinear problem

$$y'' = e^y$$

$$y(0) = 0 = y(1) .$$

This example is very easy to solve, but since it has a closed form solution it is frequently used as a test problem [8]. We have solved this problem using several of our codes, and all performed satisfactorily. In Tables V and VI, we list the results of the SUPOR Q code using interpolation and simultaneous integration, respectively. The number of derivative evaluations in Table VI indicate the sum over all iterations. The results in Table VII were obtained using the INVIMB code combined with quasilinearization and interpolation. In Table VIII, we present the results using SHOOT1. All errors were sampled at 9 equally spaced internal output points.

All of the codes performed satisfactorily except that the use of interpolation to supply the solution at the previous iteration is somewhat slower. This will become more evident in later examples. Zero values were assigned for the initial solution profile as required for the various methods.

TABLE V

SUPOR Q (Interpolation)

TOL	TIME	NIT	NIS	NDE	MAX REL ERROR-SOL
10^{-4}	0.33	3	10	NR	2.9(-09)
10^{-6}	0.35	3	11	NR	5.8(-09)
10^{-8}	0.64	4	16	NR	5.6(-09)
10^{-10}	1.15	4	31	NR	1.7(-11)

TABLE VI

SUPOR Q (Initial-Value)

TOL	TIME	NIT	NIS	NDE	MAX REL ERROR-SOL
10^{-4}	0.14	3	10	420	2.8(-09)
10^{-6}	0.26	3	20	720	6.5(-11)
10^{-8}	0.34	4	20	1040	3.5(-10)
10^{-10}	0.64	4	40	2000	6.7(-12)

TABLE VII

INVIMB (Interpolation)

TOL	TIME	NIT	NIS	NDE	MAX REL ERROR-SOL
10^{-4}	0.20	3	10	NR	6.0(-08)
10^{-6}	0.33	3	20	NR	1.8(-09)
10^{-8}	0.34	3	20	NR	6.0(-09)
10^{-10}	0.77	4	40	NR	1.5(-10)

TABLE VIII

SHOOT1

TOL	TIME	NIT	NIS	NDE	MAX REL ERROR-SOL
10^{-4}	0.06	4	10	175	8.0(-05)
10^{-6}	0.07	4	11	265	3.6(-06)
10^{-8}	0.13	5	11	663	2.5(-08)
10^{-10}	0.26	5	31	1539	1.7(-10)

Example 3:

The next problem has received much attention in the literature recently and is frequently referred to as Troesch's equation.

$$y'' = \mu \sinh \mu y$$

$$y(0) = 0,\ y(1) = 1$$

The closed form solution and tabulated values for $\mu = 10$ are listed in [4]. We have solved this problem using the SUPOR Q, INVIMB, and SHOOT4 codes. The results obtained using SUPOR Q with interpolation and simultaneous integration are listed in Tables IX and X, respectively. The interpolation procedure is now clearly having more trouble than simultaneous integration. The results obtained using INVIMB with interpolation are presented in Table XI.

The errors were computed over 29 output points which were nonuniformly distributed, being concentrated in the vicinity of $x = 1$, as in [4]. The initial guesses for SUPOR Q and INVIMB consisted of all components being identically zero over the entire interval. SUPOR Q required a couple of orthonormalization points in the beginning iterations but did not need any in the final stages of convergence. The number of derivative evaluations shown in the tables reflect the sum over all iterations.

Table XII lists the results of SHOOT4 based on using the eleven interior shooting points 0.25, 0.5., 0.7, 0.825, 0.9, 0.95, 0.975, 0.99, 0.995, 0.997, and 0.999. For this application, the absolute error tolerances used were 10^{-5} times the relative error tolerances. Although the results of SHOOT4 look fairly good on this problem it must be emphasized that the initial guesses supplied were quite good. The values used corresponded very closely to the true solution. In fact, we were unable to solve the problem using identically zero starting conditions as well as with several other "reasonable" choices. In each case, the difficulty encountered was the occurrence of a singularity in the solution of the initial-value problem within the interval of definition. This behavior makes the problem difficult for initial-value techniques. The last two lines of Table XII show the results obtained by using the Adams integrator.

This difference in the domain of convergence between the quasilinearization and the shooting approaches has been noted in several examples. In all cases, the quasilinearization procedure appeared to have a larger domain of convergence. This phenomenon merits a more careful examination which we hope to undertake.

TABLE IX

SUPOR Q (Interpolation)

TOL	TIME	NIT	NIS	NDE	MAX REL ERROR-SOL
10^{-4}	4.01	9	40	NR	1.7(-04)
10^{-6}	8.97	9	93	NR	3.3(-06)
10^{-8}	19.28	9	209	NR	4.1(-08)
10^{-10}	52.19	10	522	NR	4.2(-09)

TABLE X

SUPOR Q (Initial-Value)

TOL	TIME	NIT	NIS	NDE	MAX REL ERROR-SOL
10^{-4}	2.91	9	41	7960	1.7(-04)
10^{-6}	6.61	9	91	18476	5.7(-06)
10^{-8}	15.07	9	216	42234	3.9(-08)
10^{-10}	36.35	10	529	100462	3.8(-10)

TABLE XI

INVIMB (Interpolation)

TOL	TIME	NIT	NIS	NDE	MAX REL ERROR-SOL
10^{-4}	2.35	9	35	NR	1.7(-04)
10^{-6}	4.32	9	69	NR	1.6(-05)
10^{-8}	7.77	9	148	NR	3.2(-08)
10^{-10}	17.84	10	343	NR	3.2(-09)

TABLE XII

SHOOT4
(Eleven Shooting Points)

TOL*	TIME	NIT	NIS	NDE	MAX REL ERROR-SOL
10^{-4}	3.31	13	21	15796	8.1(-05)
10^{-6}	6.25	16	53	35602	1.8(-06)
10^{-8}	13.88	19	114	88154	1.9(-08)
10^{-10}	33.18	20	282	222355	2.0(-10)
10^{-8}	16.93	19	NR	47655	6.0(-09)
10^{-10}	23.56	20	NR	64579	1.3(-10)

*The absolute error tolerance was taken to be 10^{-5} times the relative error tolerance.

Example 4:

The next example was communicated to the authors by Ivo Babuska of the University of Maryland. The equations arose in the mathematical modeling of the kidney and show some interesting behavior. The equations are:

$$y_1' = a \frac{y_1}{y_2} (y_3 - y_1)$$

$$y_2' = - a(y_3 - y_1)$$

$$y_3' = \frac{1}{y_4}[b - c(y_3 - y_5) - ay_3(y_3 - y_1)]$$

$$y_4' = a(y_3 - y_1)$$

$$y_5' = - \frac{c}{d} (y_5 - y_3) \quad .$$

subject to the boundary conditions

$$y_1(0) = y_2(0) = y_3(0) = 1,\ y_4(0) = -10,$$

$$y_3(1) = y_5(1) \quad .$$

One of the interesting features of this example is that a good stiff integrator is required for some of the intermediate integrations of the iteration process. The values of the parameters for our experiments were $a = 100$, $b = 0.9$, $c = 1000$, and $d = 10$. The form of the boundary conditions indicates that SHOOT2 is the most appropriate of the shooting codes. As noted earlier, the shooting codes have STIFF included as one of the optional integrators whereas this feature has not been implemented in the SUPOR Q code. The stiffness is a problem only for certain values of the initial conditions, namely when $y_5(0) \leq 0.99$.

To gain some appreciation of the computational complexity of this problem, we have performed some detailed calculations. Using the variable order Adams code in [18] for a single integration of the differential equations, the following behavior was noted. With the initial value $y_5(0) = 0.99026$, the problem was relatively easy, costing about 0.88 seconds of computing time. However, with $y_5(0) = 0.99$, the execution time soars to about 101.5 seconds. The reason for this behavior is stiffness. The corresponding computing time for STIFF was 0.75 and 1.08 seconds, respectively. The code DE returns an indicator whenever 500 integration steps have been performed, and, in such circumstances, it attempts to detect whether the excessive work is due to stiffness. For the latter problem, DE reported excessive work on 26 occasions and stiffness on 81 occasions.

Other interesting features about this problem are indicated in Table XIII. There we exhibit properties of the nonlinear shooting equations $\Phi[y_5(0)] = 0$. Several points merit discussion. Firstly, there appear to be four roots of the equation, and these are rather clustered from a global point of view. This is certainly the case for the first two zeros. Secondly, the graph of Φ indicates possible difficulties for global convergence of root solvers. This is due to the extreme steepness to the left of the first zero and the flatness of the function to the right of the zeros where the residual is relatively small. Thirdly, if the iterations from the root solver stray very far to the left of the first zero, a stiff integrator will be essential to solve the problem.

In the third column of Table XIII we list the four roots, believed to be accurate to about ten digits. Apparently, only the first zero yields a physically meaningful solution because the latter three produce negative solution components for y_3 and y_5. In Table XIV, we exhibit the solution which we have obtained using SHOOT2 and $y_5(0) = 0$ as a starting guess. For tolerances 10^{-4} and 10^{-6}, the code was unable to obtain a solution because divergence of the root solver was indicated. However, for the tolerances 10^{-8} and 10^{-10} the solution given in Table XIV was found. With 10^{-8}, SHOOT2 converged in eight iterations and the computing time was 6.7 seconds. Note that the solution components vary slowly in a monotonic fashion with a maximal variation of less than a factor of two.

Because stiffness is such a problem a solution could be obtained using SUPOR Q in a reasonable amount of computing time only by giving it a guess which was significantly accurate to avoid the stiff region. Thus, with both codes we see the importance of having a stiff integrator and performing the computations quite accurately.

TABLE XIII

$\Phi[y_5(0)]=0$

$y_5(0)$	$\Phi(y_5)$	Roots of $\Phi(s) = 0 = y_5(1) - y_3(1)$
0.	6.6(+4)	
0.9	5.8(+3)	
0.99	1.4(+1)	
0.99026	4.1(-1)	
0.99027	-4.8(-2)	0.9902688359
0.99028	-1.3(-2)	
0.99029	4.6(-3)	0.9902834990
0.9903	4.0(-3)	
0.992	2.7(-5)	0.9925211341
0.993	-1.7(-5)	
1.0	-6.9(-5)	1.0304879856
1.1	1.7(-4)	
2.0	1.1(-3)	
10.0	3.4(-3)	
50.0	8.1(-3)	
100.0	1.2(-2)	

TABLE XIV

SHOOT2

x	$y_1(x)$	$y_2(x)$	$y_3(x)$	$y_4(x)$	$y_5(x)$
0.0	1.000000	1.000000	1.000000	-10.00000	0.9902688
0.1	1.080726	0.9253041	1.088425	-9.925304	1.079564
0.2	1.170692	0.8541954	1.177259	-9.854195	1.168363
0.3	1.260699	0.7932106	1.266362	-9.793211	1.257444
0.4	1.350682	0.7403668	1.355612	-9.740367	1.346684
0.5	1.440514	0.6941967	1.444836	-9.694197	1.435921
0.6	1.529858	0.6536554	1.533655	-9.653655	1.524807
0.7	1.617730	0.6181500	1.621033	-9.618150	1.612403
0.8	1.701205	0.5878186	1.703942	-9.587819	1.695978
0.9	1.771551	0.5644771	1.773387	-9.564477	1.767421
1.0	1.802778	0.5546996	1.802576	-9.554700	1.802576

Example 5:

Holt [21] considered the fifth order problem

$$\psi''' + \frac{3-n}{2}\psi\psi'' + n(\psi')^2 - 1 + G^2 - s\psi' = 0$$

$$G'' + \frac{3-n}{2}\psi G' + (n-1)\psi' G - s(G-1) = 0$$

with the boundary conditions

$$\psi(0) = \psi'(0) = G(0) = 0, \ \psi'(\infty) = 1, \ G(\infty) = 1 \ ,$$

where n and s are given constants. Problems of this form arise in the study of similarity solution for rotating flows in boundary-layer theory.

This problem is particularly sensitive for certain values of the parameters n and s. A number of authors [6, 22, 23] have solved this problem with varying degrees of success for $n = -0.1$, $s = 0.2$ and L varying from 3.5 to 11.3, where L is the interval length representing infinity. The SUPOR Q code solved the problem for these parameters quite easily. More difficult values of the parameters are for n positive and s small and positive. See [21] for an excellent discussion of the difficulties associated with this problem.

For positive values of n we started with n = 0.2, s = 0.2, and L = 200 and used the identically zero solution for all components as the initial approximation. Since we have three boundary conditions given at x = 0 and two conditions given at x = L, the most natural direction of integration would be from x = 0 to x = L. However, the intermediate solutions grow much more rapidly in the direction of increasing x than in the opposite direction. Hence, our integrations proceeded from x = L to x = 0. For larger values of n and smaller values of s, we used a simple continuation process of using the previous converged solution as the initial condition for a new set of parameters. This worked quite successfully except that small increments ($\sim$0.005) in n and s were required. The last set of parameters we used was n = 0.261 and s = 0.05, which to our knowledge has never been solved before. These results using the SUPOR Q code are outlined in Table XV and Figure 1. As a point of reference, the solution for the last set of parameters required five iterations for convergence to about five digits and the computing time was on the order of 33 seconds.

In Figure 1, notice that as n increases the amplitude of the oscillations in ψ grow and take longer to decay out. In fact, if we wish to increase n farther we should also increase L. The G function oscillates more rapidly, but the amplitude changes very little.

Holt Problems

TABLE XV

Holt Problem with n = -0.1, s = 0.2

TOL	TIME	NIT	NIS	NO	$\psi''(0)$	G'(0)	L
10^{-4}	1.79	7	28	0	-0.9900771	0.6236982	3
10^{-4}	4.35	8	58	0	-0.9663039	0.6529137	6
10^{-4}	4.85	6	88	1	-0.9663115	0.6529098	9
10^{-4}	5.24	5	118	1	-0.9663115	0.6529098	12

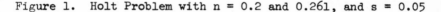

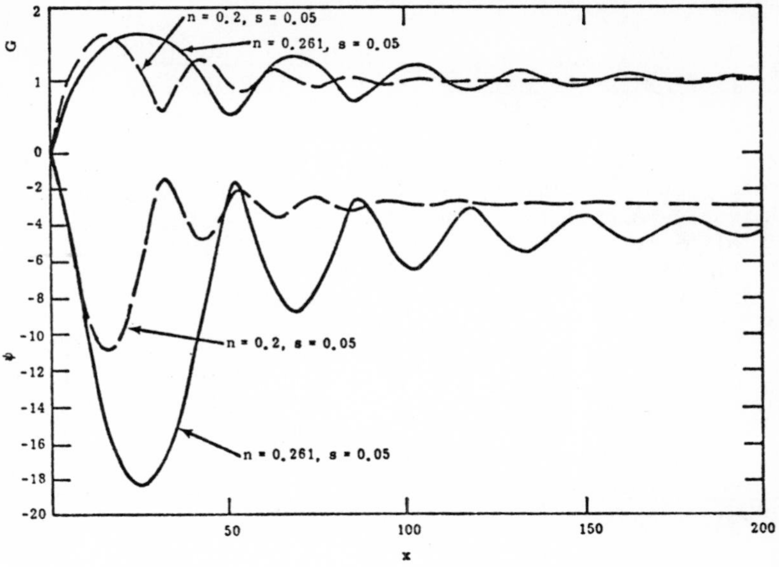

Figure 1. Holt Problem with n = 0.2 and 0.261, and s = 0.05

References

1. M. Lentini and V. Pereyra, "An Adaptive Finite Difference Solver for Nonlinear Two Point Boundary Problems with Mild Boundary Layers," to appear in SIAM J. Num. Anal.

2. R. England, N. Nichols, and J. Reid, "Subroutine DD03AD" 1973 Harwell Subroutine Library, Harwell, Berkshire, England.

3. R. Bulirsch, J. Stoer, and P. Deuflhard, "Numerical Solution of Nonlinear Two-Point Boundary-Value Problems I," to be published in Numer. Math., Handbook Series Approximation.

4. M. R. Scott and H. A. Watts, "SUPORT - A Computer Code for Two Point Boundary-Value Problems via Orthonormalization." SAND75-0198, Sandia Laboratories, Albuquerque, New Mexico 87115. Also to appear in SIAM J. Num. Anal.

5. S. Godunov, "On the Numerical Solution of Boundary-Value Problems for Systems of Ordinary Differential Equations," Uspekhi Mat. Nauk. 16 (1961), 171-174.

6. S. M. Roberts and J. S. Shipman, Two-Point Boundary-Value Problems: Shooting Methods, American Elsevier, New York, 1972.

7. M. R. Scott, Invariant Imbedding and Its Applications to Ordinary Differential Equations. An Introduction, Addison-Wesley Pub. Co., Reading, Mass. 1973.

8. M. R. Scott, "On the Conversion of Boundary-Value Problems into Stable Initial-Value Problems Via Several Invariant Imbedding Algorithms," Numerical Solution of Boundary-Value Problems for Ordinary Differential Equations. Edited by A. K. Aziz, Academic Press, N. Y., 1975. Also published as SAND74-0006, Sandia Laboratories, Albuquerque, New Mexico, 87115.

9. C. G. Broyden, "A Class of Methods for Solving Nonlinear Simulatneous Equations," Math. Comp. 19(1965), 577-593.

10. R. E. Bellman and R. E. Kalaba, Quasilinearization and Nonlinear Boundary-Value Problems, American Elsevier, New York, 1965.

11. H. B. Keller, Numerical Methods for Two-Point Boundary-Value Problems, Blaisdell Pub. Co., Waltham, Mass. 1968.

12. L. F. Shampine and M. K. Gordon, "Solving Systems of Nonlinear Equations," SAND75-0450, Sandia Laboratories, Albuquerque, New Mexico 87115.

13. R. Weiss, "The Convergence of Shooting Methods," BIT 13 (1973) 470-475.

14. E. Fehlberg, "Low-Order Classical Runge-Kutta Formulas with Step-Size Control," NASA-TA-R-315.

15. L. F. Shampine, H. A. Watts, and S. Davenport, "Solving Non-Stiff Ordinary Differential Equations - The State of the Art," SAND75-0182, Sandia Laboratories, Albuquerque, New Mexico, 87115, to appear in SIAM Review.

16. L. F. Shampine and H. A. Watts, "Practical Solution of Ordinary Differential Equations by Runge-Kutta Methods," in preparation.

17. L. F. Shampine and H. A. Watts, "Global Error Estimation for Ordinary Differential Equations," SLA-74-0198, Sandia Laboratories, Albuquerque, New Mexico, 87115. To appear in Trans. on Math. Soft.

18. L. F. Shampine and M. K. Gordon, Computer Solution of Ordinary Differential Equations: The Initial Value Problem, Freeman Press, 1975.

19. C. W. Gear, Numerical Initial Value Problems in Ordinary Differential Equations, Prentice-Hall, Inc., Englewood Cliffs, New Jersey, 1971.

20. R. D. Russell, "Collocation for Systems of Boundary-Value Problems," Numer. Math 23 (1974), 119-133.

21. J. F. Holt, "Numerical Solution of Nonlinear Two-Point Boundary Problems by Finite Difference Methods," Comm. A.C.M. 7 (1964), 366-373.

22. A. K. Agarwal, "Some Numerical Results on Holt's Two-Point Boundary-Value Problem," Aero-Astronautics Report No. 118, Rice University, 1973.

23. D. Glasser and N. DeVilliers, "Parameter Variation for the Solution of Two-Point Boundary-Value Problems and Applications in the Calculus of Variations," J. Opt. Th. Appl. 13 (1974), 164-178.

General Software for Partial Differential Equations

Niel K. Madsen and Richard F. Sincovec

1. Introduction

Over the past few years, several general purpose digital computer programs for numerically solving partial differential equations (PDEs) have been developed. These include PDEL [1], LEANS [8], DSS [11], FORSIM [2] and PDEPACK [7]. These programs all use finite difference methods for the spatial variables to convert the original space-time dependent partial differential equation system into a semi-discrete system of time dependent ordinary differential equations. Then, ordinary differential equation (ODE) techniques (i.e., the method of lines) are used to numerically solve the resulting ODE system.

With the notable exception of PDEPACK, all of these programs suffer from either the inability or lack of robustness to handle difficult stiff problems, or the inability to efficiently solve problems which require large numbers of equations and/or mesh points to achieve sufficient resolution.

Also over the past few years, many new alternatives to the use of classical finite difference methods for the numerical solution of partial differential equations have been proposed. These include variational methods, methods of weighted residuals, piecewise polynomial finite element, Galerkin and collocation methods [9], to name a few.

The basic purpose of this paper is to discuss the possible advantages of PDECOL, which is a new general purpose digital computer program for numerically solving partial differential equations. PDECOL is patterned after the reliable and efficient program PDEPACK mentioned above. The most significant feature of

Work performed under the auspices of the U.S. Energy Research and Development Administration. Contract Number W-7405-ENG-48.

PDECOL is the spatial discretization technique that is implemented. This discretization technique can best be described as a finite element collocation method which uses piecewise polynomials for the trial function space. The time integration process is accomplished by widely acceptable procedures [7] which are generalizations of the usual methods for treating time dependent partial differential equations.

2. Class of Problems

In order to develop and implement a software package based on collocation for systems of nonlinear PDEs we require the systems to be of a particular general form. The system of NPDE PDEs on the interval [a,b] must have the following structure:

$$(2.1)\quad \frac{\partial u_k}{\partial t} = f_k\left(t,x,u_1,\ \ldots,\ u_{NPDE},\ \frac{\partial u_1}{\partial x},\ \ldots,\ \frac{\partial u_{NPDE}}{\partial x},\ \frac{\partial^2 u_1}{\partial x^2},\ \ldots,\ \frac{\partial^2 u_{NPDE}}{\partial x^2}\right),$$

for $a < x < b$, $t > t_o$, and $k = 1$ to NPDE with boundary conditions

$$(2.2)\quad b_k\left(u_1,\ \ldots,\ u_{NPDE},\ \frac{\partial u_1}{\partial x},\ \ldots,\ \frac{\partial u_{NPDE}}{\partial x}\right) = z_k(t)$$

at $x = a$ and $x = b$, for $t > t_o$ and $k = 1$ to NPDE and initial conditions,

$$(2.3)\quad u_k(t_o,x) = g_k(x)$$

for $a \leqslant x \leqslant b$ and $k = 1$ to NPDE.

We assume that all of the functions, f_k, b_k, z_k, and g_k are at least piecewise continuous functions of all of their respective variables. The boundary condition functions at $x = a$ may be totally different from the corresponding functions at $x = b$. They may also differ from equation to equation. We require that the initial conditions (2.3) be consistent with the boundary conditions (2.2). "Null" boundary conditions (no condition at all) are also allowed for in PDECOL. These null conditions are useful when solving hyperbolic PDEs and ODE-PDE systems.

Equations (2.1) - (2.3) are sufficiently general to include a very wide class of physically realistic problems. Since the structure may also include many systems of PDEs for which solutions may not exist or may not be unique, it must be, of course, the user's responsibility to define a meaningful PDE problem.

3. Piecewise Polynomials

In this section we describe the piecewise polynomials which are used by PDECOL to form and evaluate a semi-discrete ODE approximation to the user's system of PDEs via a finite element collocation method.

For simplicity of presentation in the development that follows, we restrict ourselves to one PDE (i.e., NPDE = 1). PDECOL places no explicit limit on NPDE. We also assume that the user has specified a time-independent spatial mesh which consists of a sequence of NPTS points (NPTS $\geq$ 2) in [a,b] such that $a = x_1 < x_2 < \ldots < x_{NPTS} = b$. Associated with each interval in this mesh are the polynomials $p_i(x)$ for i = 1 to L, where L = NPTS - 1 is the number of spatial intervals. These polynomials form a linear space of polynomials of order K (i.e., degree < K) which we denote by P_K. Clearly, the dimension of this linear space is K * L. The mesh points $\vec{x} = (x_i)_{i=1}^{NPTS}$ will be referred to as the breakpoint sequence for P_K. Then, a piecewise polynomial function V(x), of order K, with breakpoint sequence $\vec{x}$, satisfies $V(x) = p_i(x)$ if $x_i < x < x_{i+1}$, for some polynomials $p_i(x)$, i = 1 to L.

If we impose ν continuity conditions at each of the interior breakpoints of P_K, then the dimension of this subspace of P_K, which we denote by $P_{K,\nu}$ is given by

$$N = \dim(P_{K,\nu}) = K * L - \nu * (L-1)$$
$$= K + (L-1) * (K-\nu) \quad .$$

In using PDECOL, the user has the opportunity to define the particular linear piecewise polynomial space $P_{K,\nu}$ in which the approximate solution to (2.1) - (2.3) will be determined. This requires the user to specify a spatial mesh consisting of NPTS points which will serve as breakpoints for $P_{K,\nu}$, the order K, and the number of continuity conditions, ν, at each of the interior breakpoints. We require that $3 \leq K \leq 20$, $1 < \nu < K$, and NPTS $\geq$ 2. Currently, PDECOL uses the same number of continuity conditions at each of the interior breakpoints, but it would be a trivial task to generalize to the case where different continuity conditions are permitted at different breakpoints.

A piecewise polynomial function can be represented in a variety of ways. In PDECOL we have found it convenient to determine a basis for $P_{K,\nu}$ whose elements each have minimal support (i.e., vanish outside a "small" interval). Such a basis for $P_{K,\nu}$ can be constructed by techniques described by deBoor [3]. We denote these basis functions by $\Phi_{i,K}(x)$ for i = 1 to N. This basis has the property that for any x, $a \leq x \leq b$, at most K basis functions have nonzero values. This property results in banded matrices in PDECOL as described below. Since $\{\Phi_{i,K}(x)\}_{i=1}^{N}$ forms a basis for $P_{K,\nu}$, any $V(x) \varepsilon P_{K,\nu}$ can be written as

$$(3.1) \quad V(x) = \sum_{i=1}^{N} c_i \Phi_{i,K}(x) \quad .$$

Eq. (3.1) can be used to determine the projection or approximation of an arbitrary function v(x) in the piecewise polynomial space $P_{K,\nu}$. This requires determining the coefficients c_i for i = 1 to N. To do this, we use a collocation technique by requiring Eq. (3.1) to interpolate v(x) at N collocation points, say ξ_j, for j = 1 to N. That is,

$$(3.2) \quad v(\xi_j) = V(\xi_j) = \sum_{i=1}^{N} c_i \Phi_{i,K}(\xi_j)$$

must be satisfied at the collocation points. We require that the collocation points be chosen such that $a = \xi_1 < \xi_2 < \ldots < \xi_N = b$ and $\Phi_i(\xi_i) \neq 0$ for i = 1 to N. The collocation points are selected automatically by PDECOL.

Eq. (3.2) gives a system of N linear equations whose solution gives c_i for i = 1 to N. We denote this linear system by

$$(3.3) \quad A \vec{c} = \vec{b}$$

where $A = (\Phi_i(\xi_j))$ is a banded N × N matrix with a maximum bandwidth of 2K - 1, and $\vec{b} = (v(\xi_j))$ and $\vec{c} = (c_j)$ are N-vectors.

The solution of Eq. (3.3) yields the approximation, V(x), to v(x) in the space $P_{K,\nu}$ and is given by

$$(3.4) \quad V(x) = \sum_{i=1}^{N} c_i \Phi_{i,K}(x) \quad .$$

Theoretical results [10], indicate that the maximum error between V(x) and v(x) for a < x < b is proportional to h^K where $h = \max_{2 < i < NPTS} |x_i - x_{i-1}|$. Also, Eq. (3.4) can be used to yield approximations to the first K - 1 derivatives of v(x). That is, $V^{(j)}(x)$ approximates $v^{(j)}(x)$ for j = 1 to K-1 with an error that is proportional to h^{K-j}.

We can use these approximation properties to estimate the expected orders of accuracy (in the spatial variable) of the collocation technique implemented in PDECOL. In particular, when using the piecewise polynomial space $P_{K,\nu}$, since Eq. (2.1) involves second order partial derivatives, we expect PDECOL to generate an approximate solution to (2.1) - (2.3) with errors due to spatial discretization which are proportional to h^{K-2}. For the piecewise polynomial space $P_{K,2}$, K > 3, a proper choice of collocation points [4] generates an approximate solution to (2.1) - (2.3) with errors proportional to h^K for certain classes of problems. This special case, called C^1 approximations, is the only case considered in the examples presented in Section 5.

4. Collocation Method

We now describe the finite element collocation method that is used in PDECOL to determine at any fixed time, t, a piecewise polynomial function U(t,x) in $P_{K,\nu}$ which approximates the solution

$u(t,x)$ of (2.1) - (2.3). Again for simplicity in the presentation, we consider the case for only one PDE.

Since Eq. (2.1) is time dependent, we assume the approximate solution at any time t lies in $P_{K,\nu}$ and therefore can be written in the form

$$(4.1)\quad U(t,x) = \sum_{i=1}^{N} c_i(t)\, \Phi_{i,K}(x) \quad .$$

Substituting (4.1) into (2.1) and requiring the resulting equation to be satisfied exactly at the interior collocation points, give

$$(4.2)\quad \sum_{i=1}^{N} \Phi_{i,K}(\xi_j)\, \frac{dc_i}{dt} = f(t,\xi_j,U(t,\xi_j),U_x(t,\xi_j),U_{xx}(t,\xi_j))$$

for $j = 2$ to $N - 1$. To determine the equations corresponding to $j = 1$ and $j = N$, we form special collocation equations to account for the boundary conditions (2.2). For null boundary conditions, we form an appropriate equation by using collocation on (2.1) at that boundary point in the usual manner.

Combining the boundary condition equations with (4.2) yields a system of time dependent ordinary differential equations which have the following form:

$$(4.3)\quad A\, \frac{d\vec{c}}{dt} = \vec{F}(\vec{c},t)$$

where $A = (\Phi_i(\xi_j))$ is a banded $N \times N$ matrix, $\vec{c}$ is the vector of basis function coefficients, and $\vec{F}$ is a vector with the N components

$$(f(t,\xi_j,U(t,\xi_j),U_x(t,\xi_j),U_{xx}(t,\xi_j)) \quad .$$

The first and last components of $\vec{F}$ may have been modified to account for the boundary conditions.

The initial condition, $\vec{c}(t_o)$, for Eq. (4.3) is determined by interpolating the function $g(x)$ of Eq. (2.3) in $P_{K,\nu}$. That is, we determine the c_i in Eq. (4.1) with $t = 0$ so that

$$(4.4)\quad U(t_o,\xi_j) = g(\xi_j) = \sum_{i=1}^{N} c_i(t_o)\, \Phi_{i,K}(\xi_j)$$

for $j = 1$ to N. This involves solving a linear system of equations as described in the previous section.

Eq. (4.3) can be solved using ODE techniques to integrate with respect to the time variable t. In PDECOL the time integration is performed by GEARIB, a modified version of the ODE integrator GEARB developed by Hindmarsh [5]. Modifications were made to allow for the matrix A on the left-hand side of Eq. (4.3).

5. Numerical Examples

When one considers the fact that there are over 3000 different PDE methods built into PDECOL, it should be clear that definitive and exhaustive testing of PDECOL is virtually impossible. The testing of PDECOL is currently in progress, and the results presented here are tentative.

In this section we will discuss the use of PDECOL to solve four different PDE problems which will illustrate some of the advantages and disadvantages of the use of higher order methods. Basically, our intuition tells us that for problems whose solutions are smooth and well behaved, higher order methods should "pay-off" by providing more accuracy per unit of work expended. In contrast, for more difficult problems whose solutions are not so well behaved we expect that the lower order methods will be more efficient. The results we present are graphs (at fixed times) which compare the amount of accuracy achieved by the various collocation methods with the amount of computer time required to compute the solution up to the particular fixed time. To obtain any particular data point, the following procedure was followed. First, the collocation method was determined by selecting the desired piecewise polynomial space (i.e., selecting the order K, the continuity ν, and a uniform breakpoint sequence). Then, the problem was solved using a time integration error tolerance which was small relative to the error produced by the spatial discretization. Next, this error tolerance parameter was increased on successive runs until the overall error began to increase significantly. Having thus attempted to achieve a reasonable balance between the spatial and temporal errors, the run time required was then recorded. Successive doubling of the number of mesh intervals and repeating the above procedure produced the data used to generate any given line in the graphs. In the results which follow, we will consider only the collocation methods which use the piecewise polynomials which are C^1.

Let us first consider a relatively simple PDE problem where we expect the higher order methods to be more efficient. The problem is of a simple diffusion type and is described by

$$\frac{\partial u}{\partial t} = \frac{\partial^2 u}{\partial x^2} + \pi^2 \sin \pi x \quad ,$$

$$u(0,x) = 1 \quad ,$$

$$u(t,0) = u(t,1) = 1 \quad ,$$

for $0 < x < 1$ and $t > 0$. The exact solution for this problem is easily seen to be

$$u(t,x) = 1 + (\sin \pi x)\left(1 - e^{-\pi^2 t}\right) \quad .$$

Shown in Figure 1 is a graph of the maximum errors in the computed solutions at t = 0.1 for various piecewise polynomial orders versus the computer times (CDC-7600) required to produce the solution. For comparison purposes, results produced by the second order finite difference code PDEPACK are shown. Since greater accuracies require more mesh points, and hence result in larger numbers of equations (4.2) to be solved, longer run times are, of course, to be expected. When greater accuracies are demanded, the results show that for this problem, the collocation methods are uniformly better than the finite difference results. Also, as expected, for greater accuracies (10^{-3} or more) the higher order methods are consistently more efficient.

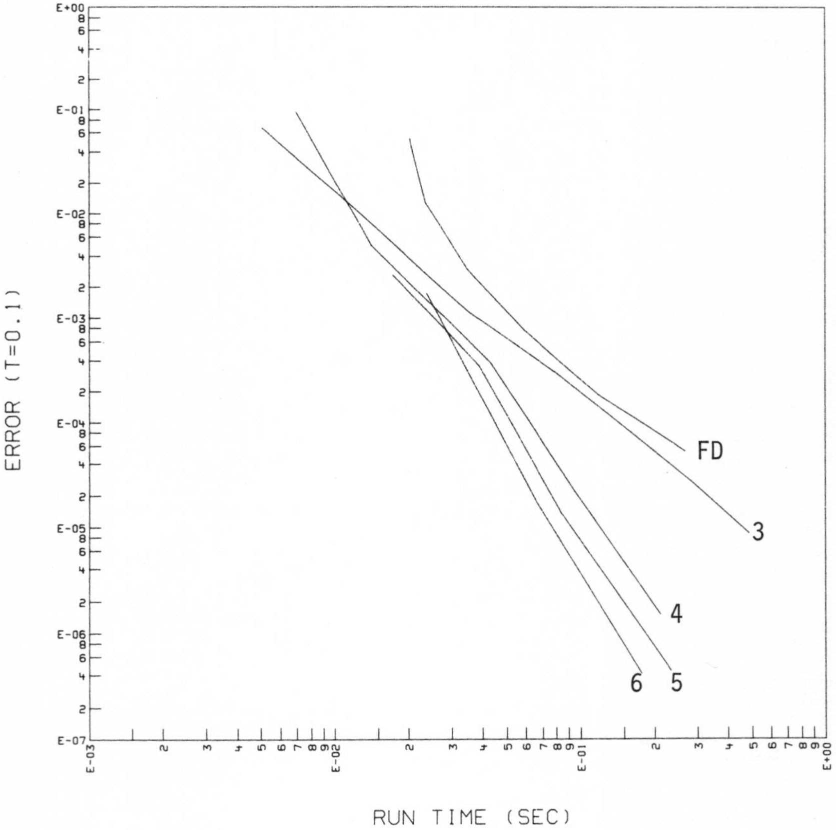

Fig. 1. Simple diffusion problem: maximum errors in the approximate solutions obtained by PDECOL using C^1 piecewise polynomials of orders 3-6 at t = 0.1 are plotted versus the required run times. Analogous second order finite difference results (FD) are also shown.

Next, we consider the more difficult and nonlinear Burgers equation problem given by

$$\frac{\partial u}{\partial t} = \nu \frac{\partial^2 u}{\partial x^2} - u \frac{\partial u}{\partial x} \quad ,$$

$$u(0,x) = \Phi(0,x) \quad ,$$

$$u(t,0) = \Phi(t,0) \quad ,$$

$$u(t,1) = \Phi(t,1) \quad ,$$

where

$$\Phi(t,x) = \frac{.1e^{-A} + .5e^{-B} + e^{-C}}{e^{-A} + e^{-B} + e^{-C}}$$

and

$$A = \frac{.05}{\nu}(x - .5 + 4.95t) \quad ,$$

$$B = \frac{.25}{\nu}(x - .5 + 0.75t) \quad ,$$

$$C = \frac{.5}{\nu}(x - .375) \quad .$$

The function $\Phi(t,x)$ is the exact solution for this problem. This problem represents a slightly diffusive nonlinear wave moving from left to right on [0,1]. For small values of ν the problem becomes very difficult as a steep shock-like wave front develops. We consider this problem for $\nu = 0.003$. The time evolution of the solution is shown in Figure 2.

Figure 3 shows the performance of PDECOL for this Burgers equation problem. Again, we note that the collocation results are uniformly better than the finite difference results. However, for this problem the results obtained using the C^1 piecewise polynomial spaces of orders 4, 5 and 6 are very similar, with the higher order results only slightly better for the range of errors less than about 4.E-03.

As a problem of a different type, we consider the linear hyperbolic wave propagation problem given by

$$\frac{\partial u}{\partial t} = - \frac{\partial u}{\partial x} \quad ,$$

$$u(t,0) = 1 + .5 \sin(-4\pi t) \quad ,$$

$$u(t,1) = \text{Null boundary condition} \quad ,$$

$$u(0,x) = 1 + .5 \sin(4\pi x) \quad .$$

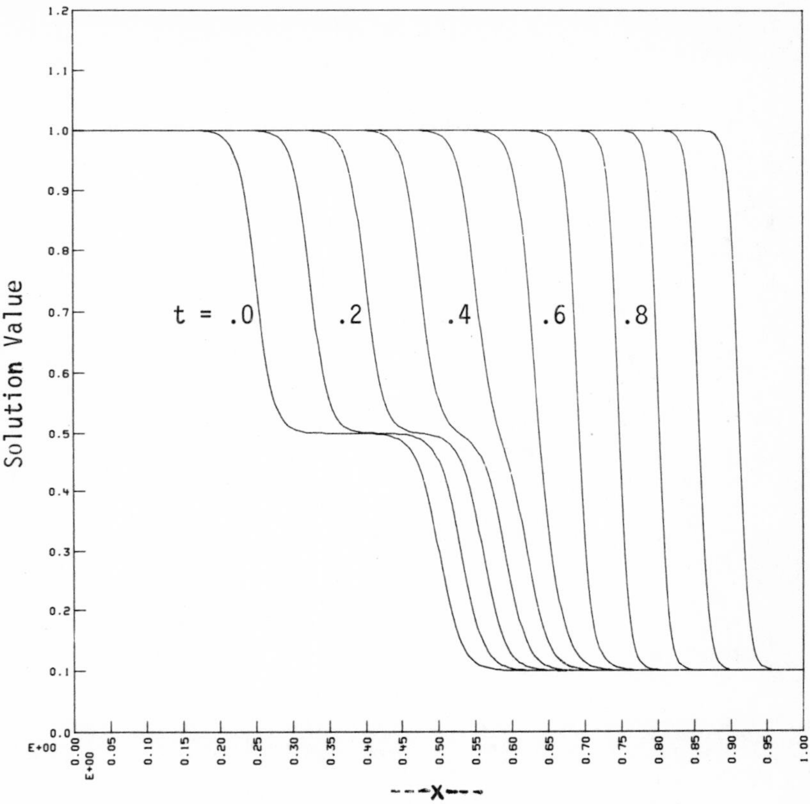

Fig. 2. Nonlinear Burgers problem: time evolution of the exact solution.

Clearly, the exact solution on [0,1] is given by

$$u(t,x) = 1 + .5 \sin(4\pi(x-t)) \quad .$$

This problem illustrates the use of PDECOL for a hyperbolic problem where a null boundary condition exists. The performance of PDECOL is shown in Figure 4. We note that for requests of low accuracies (1.E-01 to 5.E-04), the higher order methods perform about equally well and again much better than the finite difference method and the collocation method with polynomials of order three. For the higher accuracies, the efficiency of high order methods is readily apparent.

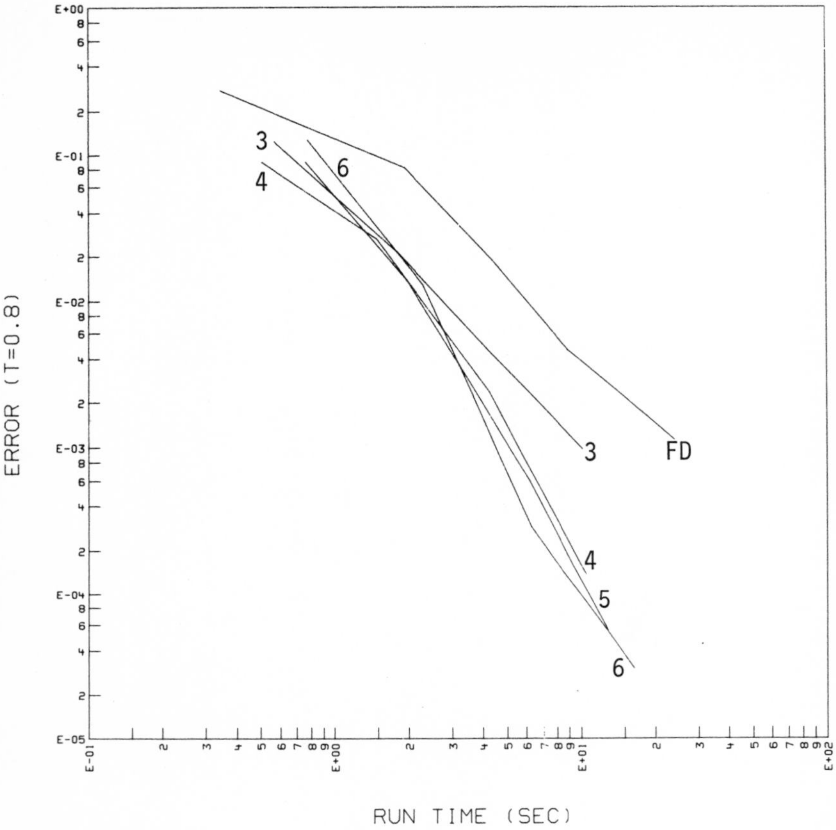

Fig. 3. Nonlinear Burgers problem: maximum errors in the approximate solutions obtained by PDECOL using C^1 piecewise polynomials of orders 3-6 at t = 0.8 are plotted versus the required run times. Finite difference results (FD) are also shown.

As a concluding illustrative example, we consider the system of partial differential equations

$$\frac{\partial u}{\partial t} = \frac{\partial}{\partial x}\left((v-1)\frac{\partial u}{\partial x}\right) + [16xt - 2t - 16(v-1)](u-1) + 10xe^{-4x}$$

$$\frac{\partial v}{\partial t} = \frac{\partial^2 v}{\partial x^2} + \frac{\partial u}{\partial x} + 4u - 4 + x^2 - 2t - 10te^{-4x}$$

with boundary conditions

$$u(t,0) = v(t,0) = 1.0 \qquad \text{at} \quad x = 0,$$

$$3u + \frac{\partial u}{\partial x} = 3 \quad \text{and} \quad 5\frac{\partial v}{\partial x} = e^4(u-1) \qquad \text{at} \quad x = 1,$$

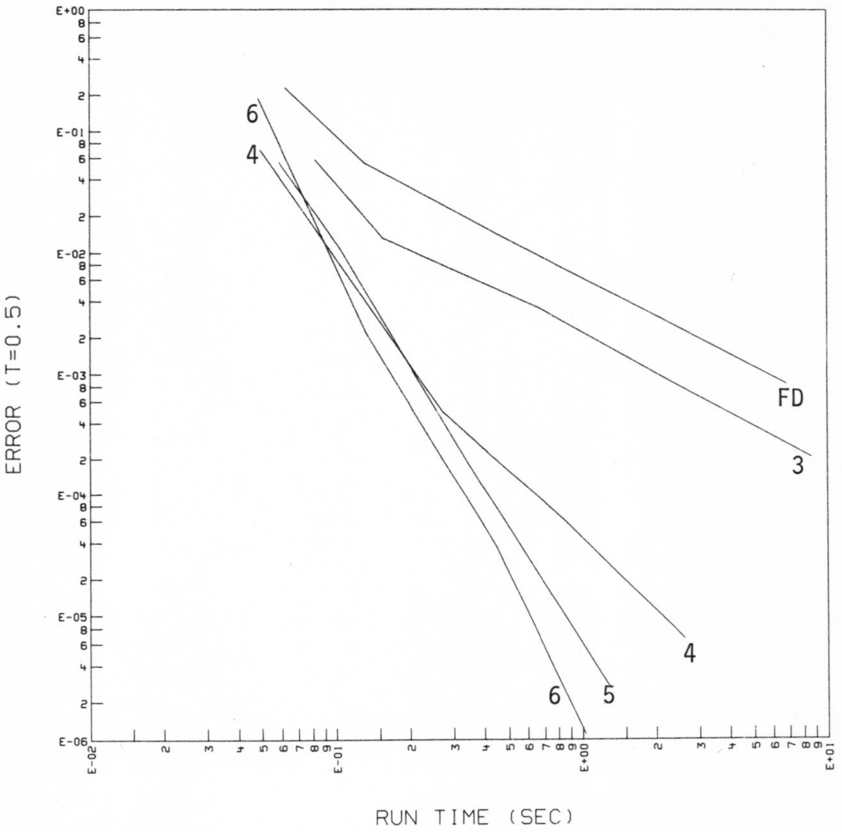

Fig. 4. Hyperbolic wave propagation: maximum errors in the approximate solutions obtained by PDECOL using C^1 piecewise polynomials of orders 3-6 at t = 0.5 are plotted versus the required run times. Analogous second order finite difference results (FD) are also shown.

and initial conditions

$$u(0,x) = v(0,x) = 1 \quad .$$

The exact solution for this problem is

$$u(t,x) = 1 + 10xte^{-4x}$$

$$v(t,x) = 1 + x^2t \quad .$$

Figures 5 and 6 show the errors in the two solution components produced by PDECOL versus the computer time required. Here we see that for error levels in the range of 0.1 to 0.001 the

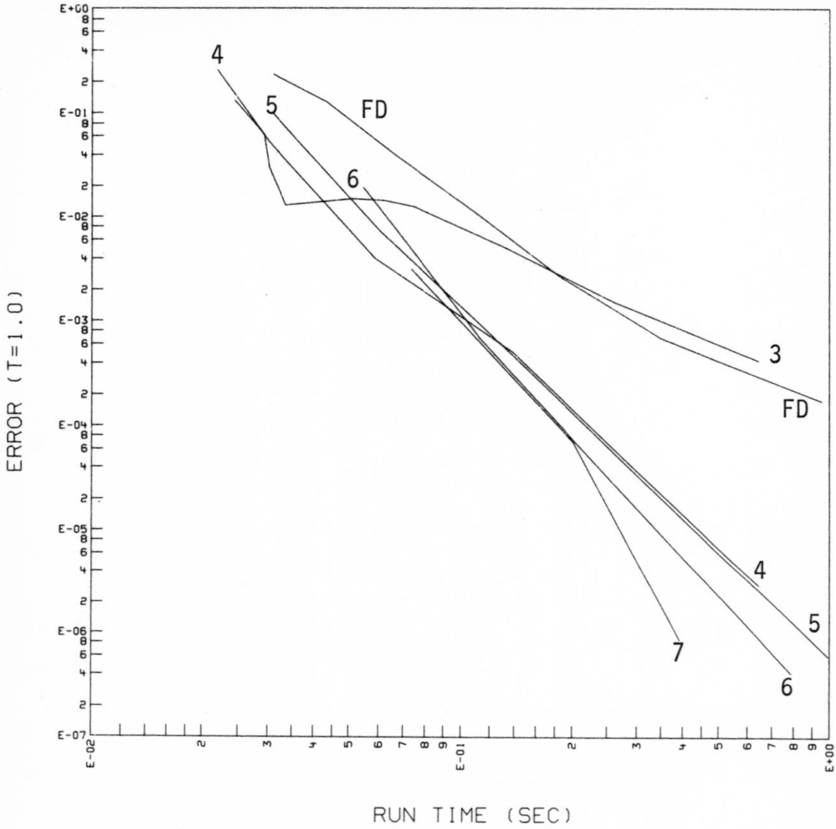

Fig. 5. Coupled system problem, first solution component u: maximum errors in the approximate solutions obtained by PDECOL using C^1 piecewise polynomials of orders 3-7 at t = 1.0 are plotted versus the required run times. Analogous second order finite difference results (FD) are also shown.

lower order collocation methods are the most efficient, and for the range below 0.0001 or 0.00001, the higher order methods become the most efficient. Since the solution to this problem has such a trivial dependence upon the time t, a slightly different procedure was used to obtain these results. For this problem, the time integration procedure will obtain essentially the exact solution to the equations (4.2) regardless of what time integration error tolerance is specified. To obtain results which we feel would be consistent with those for less trivial problems, we set the time integration error tolerance to be a factor of ten smaller than the maximum error produced in the second solution component. This roughly corresponds to the values which were required for the other problems which had less trivial behavior.

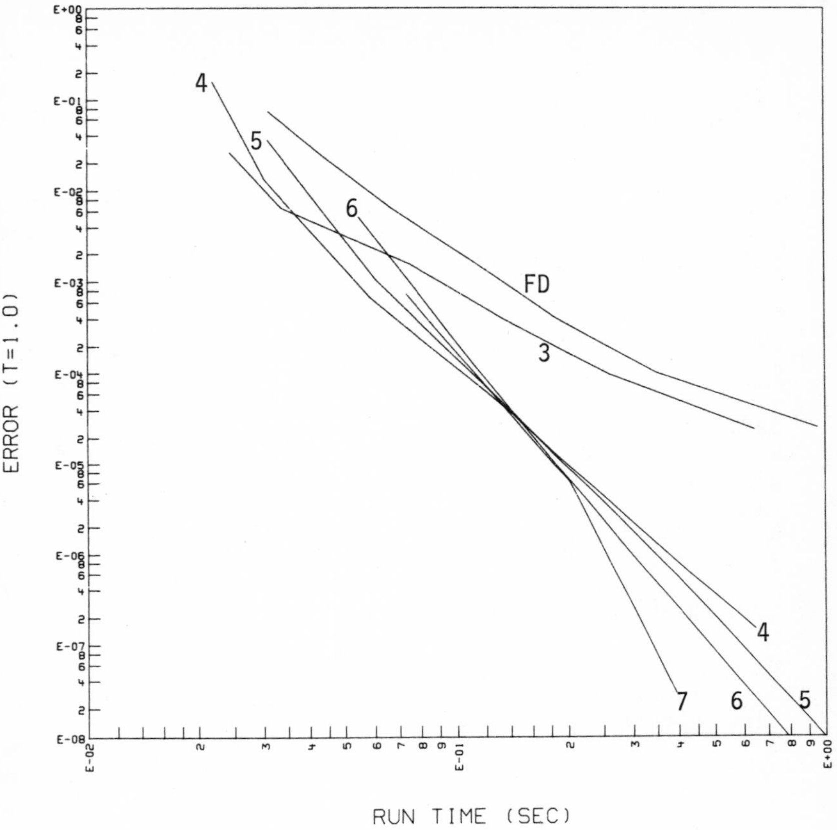

Fig. 6. Coupled system problem, second solution component v: maximum errors in the approximate solutions obtained by PDECOL using C^1 piecewise polynomials of orders 3-7 at $t = 1.0$ are plotted versus the required run times. Analogous second order finite difference results (FD) are also shown.

6. Conclusions

Our experience in using PDECOL has led us to believe firmly that PDECOL is an extremely versatile and unique software package. It can reliably solve a very broad class of nontrivial and nonlinear systems of partial differential equations. Its higher order methods can produce extremely accurate solutions quite efficiently when compared to lower order methods. The package is quite portable and very easy to use.

REFERENCES

[1] Cardenas, A.F. and W.J. Karplus, "PDEL - A Language for Partial Differential Equations," Comm. ACM, Vol. 13, No. 3, March 1970, pp. 184-191.

[2] Carver, M.B., "FORSIM: A Fortran Oriented Simulation Package for the Automated Solution of Partial and Ordinary Differential Equation Systems," Report AECL-4608, Atomic Energy of Canada, Limited, Chalk River, Ontario, 1973.

[3] de Boor, C., "On Calculating with B-Splines," J. Approx. Theory, Vol. 6, No. 1, July 1972, pp. 50-62.

[4] de Boor, C. and B. Swartz, "Collocation at Gaussian Points," SIAM J. Numer. Anal., Vol. 10, No. 4, September 1973, pp. 582-606.

[5] Hindmarsh, A.C., "Preliminary Documentation of GEARIB: Solution of Implicit Systems of Ordinary Differential Equations with Banded Jacobian," Lawrence Livermore Laboratory, UCID-30130, February 1976.

[6] Madsen, N.K. and R.F. Sincovec, "PDEPACK: A New Tool for Simulation," Proceedings of the 1975 Summer Computer Simulation Conference, San Francisco, CA, July 1975.

[7] PDEPACK: Partial Differential Equations Package Reference Manual, Scientific Computing Consulting Services, P.O. Box 335, Manhattan, KS 66502, 1975.

[8] Schiesser, W.E., "A Digital Simulation System for Mixed Ordinary/Partial Differential Equation Models," Proceedings of IFAC Symposium on Digital Simulation of Continuous Processors, Gyor, Hungary, September 1971, Vol. 2, pp. S2-1 to S2-9.

[9] Strang, G. and G.J. Fix, An Analysis of the Finite Element Method, Prentice-Hall, 1973.

[10] Swartz, B.K. and R.S. Varga, "Error Bounds for Spline and L-Spline Interpolation," J. Approx. Theory, Vol. 6, No. 1, July 1972, pp. 6-49.

[11] Zellner, M.G., "DSS-Distributed System Simulator," Ph.D. Thesis, Lehigh University, Bethlehem, PA, 1970.

The Choice of Algorithms in Automated Method of Lines Solution of Partial Differential Equations

M. B. Carver

INTRODUCTION

The method of lines is a general approach to the solution of partial differential equations, in which spatial derivatives are approximated by appropriate finite difference formulae. This converts each partial differential equation (PDE) into a set of coupled ordinary equations (ODE's) in time, which may then be integrated. Early applications used simple difference formulae for the spatial derivatives and integrated the resulting ODE's continuously on an analogue computer.

The development of error controlled variable step size integration algorithms which permit a digital computer to approximate continuous integration, led to a resurgence of interest in the method of lines. Because of its inherent generality, the method has been popular in programs designed to automate the solution of PDE's[1-6].

These programs operate on PDE's supplied by the user and automatically perform the conversion to ODE's and subsequent integration, according to clearly defined internal procedures. While the use of such a proven program will produce results faster and more reliably than ad hoc programming, the accuracy of these results remains difficult to assess.

Accuracy is, of course, dependent on both the integration algorithm and the method of spatial discretization. However, concentrated effort in the field of integration algorithms has advanced the science sufficiently that, providing a proven algorithm is used with a realistic error bound, the effect of integration inaccuracies is small compared to those introduced by discretization. The efficiency of the integration algorithm used, however, is of paramount importance, particularly as the resulting ODE's frequently constitute a large stiff system which cannot easily be solved by normal methods.

Improvements in accuracy may always be obtained by using more divisions in the spatial grid, but this requires a disproportionate amount of computing time as it frequently reduces the permissible time step. Recent investigations show that using high order discretization formulae greatly increases accuracy at the expense of little extra computing time[4,5], but unfortunately this simple remedy does not apply to every class of equation.

The problem has further degrees of freedom, as one may develop discretization formulae from any valid interpolation polynomial, use equal or unequal spatial divisions and choose between central, biased or directional derivative formulae of any order.

This paper is based on experiences gleaned while applying a particular method of lines PDE simulation package, FORSIM[6], to a wide spectrum of academic and engineering problems, but the comments made have general import.

An automated PDE package must cater to two kinds of users. The first requires a quick solution to a set of equations he is studying by themselves, or as an isolated section of a more complex analysis. He is interested in obtaining results with a minimum personal effort, and is not concerned with accuracy or economy of computing resources. He may even be annoyed if required to read the manual.

The second type of user is developing a large simulation on which he plans a large number of production runs. He demands that the automated system be asier to use, faster, and more accurate than programs written specially for his particular application, but he is willing to spend some time on his implementation in order to achieve these goals.

Like most software of its kind, FORSIM is extremely robust, and provided that user number one observes the ground rules, he is guaranteed a solution of some sort. Although user number two may not fully realize all his goals, he is provided with a number of options which permit him to optimize the accuracy and efficiency of his implementation. To minimize overhead, these options are controlled and selected by the user, rather than by the package. This paper attempts to provide rudimentary guidelines to the use of these options by those who are interested in the accuracy and efficiency of their solutions.

Concerning the choice of integration algorithm, some thoughts are offered on the relative merits of the simple Euler formula at one extreme, as compared to the recently developed sparse matrix version(7) of the Hindmarsh-Gear algorithm(8). For spatial discretization, it discusses if, when and how, unequal spatial division should be used, and the choice of order, type and directionality of spatial derivative formulae. Finally it mentions how large simulations may be developed from the basic package and then increased in efficiency by short circuiting some of the features of the automated system.

THE FORSIM PDE PACKAGE

The FORSIM package is designed with both the above classes of user in mind. The PDE's or ODE's are written by the user in a simple FORTRAN subroutine according to clearly defined rules. This routine has several sections which define initial conditions, boundary conditions, printout, and the equations themselves. This is done in several separate routines in most other software packages(1-4). While the two approaches are about equal in user convenience, the FORSIM method offers a distinct advantage in computational efficiency. All evaluation of derivatives, definition of equations and application of boundary conditions may be done together with DO loops for all NE=NPDE*NPOINT equations, whereas separate routines must be called NE times, thus requiring a considerable amount of overhead.

THE INTEGRATION ALGORITHM

The algorithm must integrate the system of ODE's

$$[Y'] = [f(Y)] = [J][Y] \tag{1}$$

A solution is normally completed in two phases. In the debug runs, all that is required is an indication of whether the calculation is progressing in an encouraging manner. A sophisticated integration algorithm may cloud the issue here, and the simplest fixed step Euler algorithm will suffice. Once it is established that the problem is defined correctly, an error controlled algorithm should be used to obtain the final solution.

THE HINDMARSH-GEAR ALGORITHM

Although a number of integration routines have been available in FORSIM, including Runge-Kutta, Fowler-Warten, Adams, Gear and Bulirsch-Stoer, the implementation of Gear's algorithm(8) by Hindmarsh(9) has proved the most robust. This requires less storage and time for matrix manipulation than the original Gear.

The power of the Gear algorithm lies in the combination of implicit backward differentiation formulae and Newton iteration, which results in the following expression for the m+1th estimation of Y at time t_n

$$Y_{n(m+1)} = Y_{n(m)} - P^{-1}_{n(m)} \phi(Y_{n(m)}) \tag{2}$$

where $$P_n = I - h\beta_0 J_n$$

and $$J_n = \frac{\partial F}{\partial Y}\Big|_{Y_{n(m)}}$$

An accurate representation of the Jacobian matrix J permits the iteration 2 to converge rapidly for large step size h. Hindmarsh presents several options in which the Jacobian can be taken as a full, banded, diagonal or unit matrix. For a general system the above list is stated in order of decreasing accuracy and storage requirement (if the matrix is truly banded, the second option is fully accurate).

Recent work on the neutron kinetics equations[10] showed that satisfactory solutions could be obtained only from the full matrix option, but the storage and time required to manipulate the Jacobian became prohibitive for large systems. This motivated the use of sparse matrix techniques to handle equation (2).

THE SPARSE MATRIX OPTION WITH JACOBIAN EVALUATION OPTIMIZATION

The integration routine now recommended to FORSIM users combines the Hindmarsh algorithm with the sparse matrix routines of J.K. Reid[11].

Apart from the obvious storage advantage, the most important feature of the new algorithm is the optimization of the Jacobian evaluation procedure. The most efficient means of obtaining J is by means of a special user-provided subroutine which defines the NE^2 elements of the matrix. However, for general simulations this is impractical, and Gear provided an option in which each of the variables Y_j is perturbed and the effect on the functions F_i then determined. This requires NE calls to the function routine. The new algorithm first assesses the sparsity structure of the Jacobian and determines how many of the Y_j can be perturbed at one time only influencing one F_i. This greatly reduces the number of function calls. In a test case of 242 equations, only 7 calls were required as opposed to 242 previously. Some figures illustrating the performance of the new algorithm are given in Table 1.

TABLE 1

Gear's Algorithm Performance

Method	Functional Iteration	Diagonal Matrix	Banded Matrix	Full Matrix	Sparse Matrix	4th Order Runge-Kutta
Jacobian Storage	0	242	1,000	60,000	10,000	0
No. of Function Calls to Evaluate Jacobian	0	1	242	242	7	0
At 4×10^{-5} secs						
No. Steps	269	60	9	9	9	110
No. Function Calls	1,098	384	983	983	281	1,475
CP Time	35.	10.	18.	75.	4.	29.
At 10^{-3} secs						
No. Steps	>3,000	768	32	15	15	-Incomplete-
No. Function Calls	>12,000	5,960	5,660	1,480	306	
CP Time	>300.	155.	100.	228.	6.	
At 10^{-1} secs						
No. Steps	------------ Incomplete ------------			75	75	
No. Function Calls				3,728	452	
CP Time	10^{5}*	3×10^{4}*	10^{4}*	375.	35.	10^{5}*
Max. Time Step	10^{-7}	10^{-6}	10^{-4}	10^{-2}	10^{-2}	10^{-7}

*Estimated

THE DISCRETIZATION FORMULAE FOR SPATIAL DERIVATIVES

To obtain the solution of partial differential equations of the form

$$U_t = \phi(x,t,U,U_x,U_{xx},V,V_x,V_{xx},\ldots)$$
$$V_t = \psi(x,t,U,U_x,U_{xx},V,V_x,V_{xx},\ldots) \qquad (3)$$

the variables U and V are approximated by arrays of NPT discrete points of space, and their spatial derivatives are obtained by differentiating any suitable approximation polynomial the requisite number of times. A general approximation formula may take into account any number of equally or unequally distributed points. There are several well-known possibilities.

LAGRANGE FORMULAE

The approximation of u(x) based on n surrounding discrete points $u(x_i)$ is

$$u(x) = \sum_{i=1}^{n} L_i(x)\ u(x_i) \qquad (4)$$

where L_i is the Lagrange polynomial of any order, n-1, where $n \leq$ NPT.

$$L_i = \frac{\prod_{j=1(\neq i)}^{n} (x-x_j)}{\prod_{j=1(\neq i)}^{n} (x_i-x_j)}$$

Thus

$$u_x(x) = \sum_{i=1}^{n} L_{ix}(x)\ u(x_i) \tag{5}$$

$$u_{xx}(x) = \sum_{i=1}^{n} L_{ixx}(x)\ u(x_i) \tag{6}$$

These formulae require only the discrete functional values $u(x_i)$ and are equivalent to using an n point Lagrange polynomial at each of the NPT points. Although none of these polynomials is related to its neighbour, other than through the values $u(x_i)$, accurate results can be obtained.

HERMITE FORMULAE

Improved accuracy can be obtained by also using the derivative values $u_x(x_i)$ in the Hermite approximation

$$u(x) = \sum_{i=1}^{n} (h_{ij}(x)u(x_i) + h_{2i}(x)u_x(x_i)) \tag{7}$$

One may, therefore, couple neighbouring polynomials through u_x, carrying a continuous solution of u_x as well as u by deriving a further equation from (3)

$$u_t = \phi(x,t,u,u_x,u_{xx},\ldots)$$

$$\frac{\partial u_t}{\partial x} = \frac{\partial t}{\partial x} [\phi(x,t,\ldots)]$$

Hence,

$$\frac{\partial}{\partial t} (u_x) = \frac{\partial}{\partial x} [\phi(x,t,u,\ldots)] \tag{8}$$

The additional differentiation may generate third derivatives, and the formulae required for u_{xx} and u_{xxx} are obtained by differentiating (7).

Further refinements to this technique involve higher orders of derivatives in the interpolation formulae and thus further equations must be integrated[13].

OTHER TECHNIQUES

Another possible way of obtaining derivatives is to use spline functions which fit piecewise nth degree polynomials to the functions and produce continuous derivatives up to order n-1. These have been found to give very accurate approximations to functions and their derivatives. Legendre polynomials have also been used[14].

Finally, there has been increasing interest in applying the method of lines in conjunction with techniques traditionally associated with finite elements. Variational and weighted residual methods, such as Galerkin and collocation techniques have been very successful in particular applications. In general, these extend the above techniques by representing the function u in the differential equation

$$\vec{u}_t - \vec{\phi}(\vec{u},t) = 0 \tag{9}$$

by an approximation function $\vec{u}_a$, and minimizing the resulting residuals. This normally converts the equations (9) into a related set

$$\vec{A}\,\vec{\alpha}_t - \vec{\beta}(\vec{\alpha},t) = 0 \tag{10}$$

where A, α and β are determined from u, u_a, ϕ and equations (9) in a manner dictated by the technique used. This introduces considerably more complexity into the solution but requires fewer points for a given accuracy[15-17].

IMPLEMENTATION

Although the more complex techniques pay dividends in accuracy, the simplicity of the Lagrange approach has popular appeal, and this technique is used in several general PDE solution packages[1,3-6]. To alert users of such packages, the strengths and weaknesses of this approach will be discussed.

The symmetric nature of the Lagrangian formulae permits general routines to be written for interpolation or differentiation which will use any number of equally or unequally spaced points[12], but such general evaluation is again time consuming. It is computationally more efficient to derive and use specific formulae for each order of approximation, although one is then limited by the total number of formulae he cares to derive and code. If the analysis is restricted to equally spaced points, some generality can be regained by merely storing the necessary coefficient vectors[18].

The subsequent discussion assumes that routines to evaluate spatial derivatives are available for equal and unequal spatial divisions, and offers guidelines in the choice of order of approximation and directionality.

UNEQUAL SPATIAL DIVISION

Unequal spatial division should be used only when the nature of the problem makes it unavoidable, as computational disadvantages well outnumber any practical advantage. The order of the error is lower for unequal divisions, the calculation of the spatial derivatives is considerably more time consuming, integration is slow, as the time step of the integration routine is limited by the smallest spatial increment rather than the number of divisions, and the choice of unequal division cannot be done arbitrarily or further errors are introduced.

There can be a case for using unequal divisions when the system under consideration has a small stationary area within which most of the steep spatial gradients are confined. For example, a long coolant pipe with a small heated section would satisfy this criterion. Even in this case, the spatial division must not be dictated entirely by the geometry, but should be established as a smooth variation such that neighbouring increments do not differ greatly in size (i.e. dx(x) should be differentiable[14]. Defining dx(x) as a smooth function has the added advantage that the spatial grid can be calculated rather than entered in an array.

ERROR AND THE SPATIAL DERIVATIVE APPROXIMATION

The most widely used formulae are the three-point symmetric analogs for equal increments h in x which may be derived from (5) and (6):

$$u_{xi} = (u_{i+1} - u_{i-1})/2h \tag{11}$$

$$u_{xxi} = (u_{i+1} - 2u_i + u_{i-1})/h^2 \tag{12}$$

These may be applied throughout the spatial range except at the two end points, which may be handled by assymmetric formulae or by the establishment of 'pseudo points' beyond the boundary.

The order of the truncation errors in (11) and (12) is

$$e_{ux} \sim \frac{1}{6} h^2 u^{(3)}$$

$$e_{uxx} \sim \frac{1}{12} h^2 u^{(4)}$$

The truncation error of the symmetric formula may be reduced markedly by using, for example, 5-point formulae:

$$e_{ux_5} \sim \frac{1}{30} h^4 u^{(5)}$$

$$e_{uxx_5} \sim \frac{1}{90} h^4 u^{(6)}$$

and similar advantage is gained with higher orders.

However, with m-point difference formulae, derivatives at (m-1)/2 points either end must be calculated by skewed difference formulae, and the truncation error here can be an order of magnitude higher. For example, the errors of the second derivatives at the first two points using a 5-point scheme are

$$\text{(i)} \quad e_{uxx_5} \sim \frac{5}{6} h^3 u^{(5)}$$

$$\text{(ii)} \quad e_{uxx_5} \sim \frac{1}{12} h^3 u^{(5)}$$

The reduced rate of improvement in accuracy from the use of skewed formulae near the end points is illustrated clearly in Figure 1. If formulae incorporating an equal number of points throughout the grid are used, error tends to build up near the boundaries. When the initial transient is large, errors may be negligible, but in problems in which initial conditions represent an equilibrium state, these errors can generate completely spurious transients.

A scheme for attaining increased accuracy through the use of higher order formulae must, therefore, ensure that the order of the truncation error, rather than the order of formula is kept uniform across the spatial grid.

This problem is further compounded if the function to be differentiated yields continuous derivatives of increasing magnitude. For example, the function u = log x gives derivatives

$$u^{(n)} = (-1)^{n-1}(n-1)!x^{-n}$$

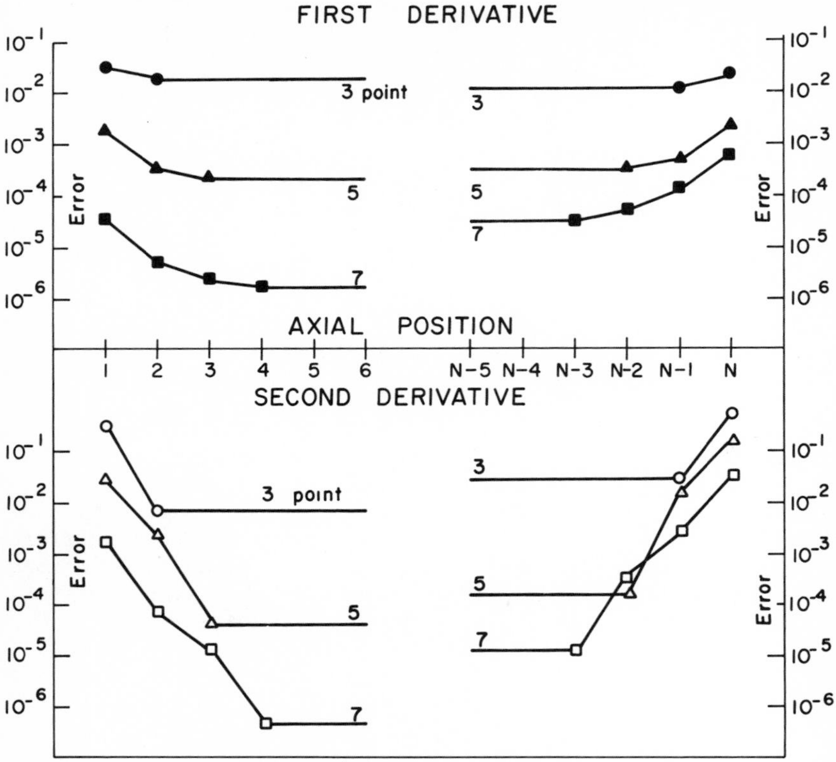

Fig. 1. Relative Error in Symmetric Difference Formulae for Two Functions. u = sin πx, dx = 0.1, u = log x

Since the accuracy of the skewed formulae increases less rapidly with order than the symmetric formulae, the $(n-1)!$ factor can almost eliminate the advantage of using higher orders. This is also illustrated in Figure 1.

Finally, the high order centered difference scheme is not necessarily optimal for all problems. In the analysis of hyperbolic PDE's, directional derivatives (forward, backward or biased) are frequently more accurate. This is discussed under Upwind Differencing below.

PARABOLIC AND ELLIPTIC EQUATIONS

Provided the above reservations are kept in mind, good results can be obtained from high order symmetric formulae, particularly for parabolic equations. Consider, for example, a problem involving Burger's equation discussed by Madsen and Sincovec(3):

$$\frac{\partial u}{\partial t} = - \frac{\partial u}{\partial x} + \frac{\partial^2 u}{\partial x^2} \tag{13}$$

with solution

$$u(t,x) = \frac{.1e^{-A} + .5e^{-B} + e^{-C}}{e^{-A} + e^{-B} + e^{-C}} \tag{14}$$

where

$$A = \frac{50}{3} (x - .5 + 4.95t)$$

$$B = \frac{250}{3} (x - .5 + .075t)$$

$$C = \frac{500}{3} (x - .375)$$

The solution represents one shock wave catching up and merging with another. The paper gives results from runs with three-point difference formulae with 200 and 400 points.

The error reported was verified in FORSIM, indicating that the method of lines solution is reproducible, even though the two software packages appear quite different to the user. However, as shown in Figure 2, higher order formulae permit a large reduction in the number of points for the same accuracy, thus giving much lower computer times.

Elliptic equations in space may be converted to parabolic equations in space and time by adding a time derivative term, and integrating until this becomes effectively zero throughout the range[3], or by integrating along a coordinate and iterating on the boundary condition[19].

HYPERBOLIC EQUATIONS

Hyperbolic equations may also be solved by using symmetric difference formulae, but the probability of getting a good solution is considerably lower. Consider the simplest first order hyperbolic equation, the one-dimensional advective equation representing flow of fluid without friction or heat transfer

$$\frac{\partial u}{\partial t} = -c \frac{\partial u}{\partial x} \tag{15}$$

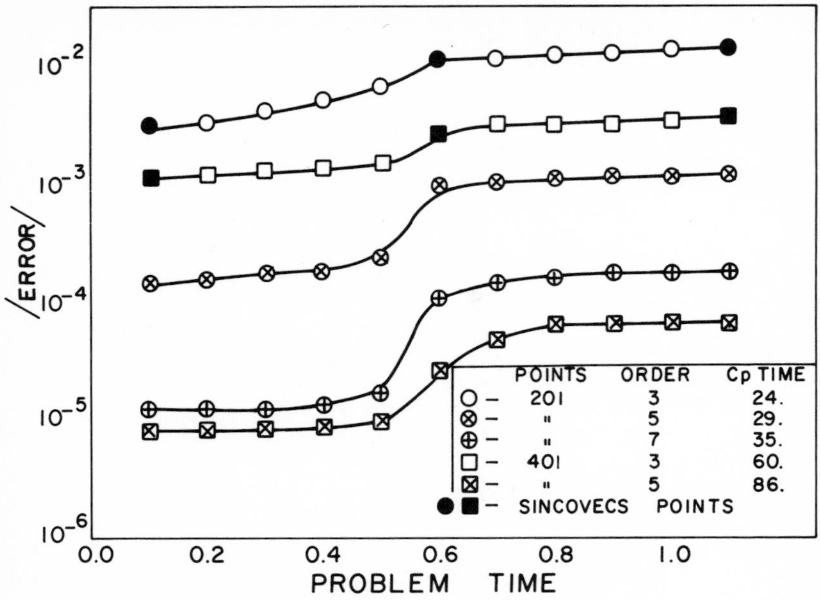

Fig. 2. Errors in Method of Lines Solution of Burger's Equation.

It can be shown[20] that the application of the three-point symmetric approximation to $\partial u/\partial x$ gives arise to forward and backward propagating waves consistent with the solution of the wave equation.

$$\frac{\partial^2 u}{\partial t^2} = -c \frac{\partial^2 u}{\partial x^2} \tag{16}$$

In the true solution of (15) the initial disturbance propagates forwards, entirely without attenuation. The spurious backward wave can be magnified by reflection at the boundaries until it eventually dominates the solution. This spurious wave problem is common to all hyperbolic equation systems, and for this reason such systems should be approached with care. Detailed knowledge of their behaviour is probably only to be gained by an extensive study of the method of characteristics, but the difficulty of applying this method to systems of several dependent variables has motivated the development of suitable finite difference techniques.

Luckily, some of these techniques can be easily applied within the frame of reference of the generalized method of lines approach.

ARTIFICIAL DISSIPATION

Frictionless flow, of course, is somewhat artificial, and the addition of a dissipative term to (15) not only helps the numerical solution by converting it to a parabolic equation and damping out the spurious oscillations, but is also physically more realistic. The equation then becomes

$$\frac{\partial u}{\partial y} = -c\,\frac{\partial u}{\partial x} + \alpha\,\frac{\partial^2 u}{\partial x^2} \tag{17}$$

Madsen and Sincovec[3] illustrate the use of artificial dissipative terms to overcome the shock in the hydraulic jump problem, but choice of α is not arbitrary and must be related in some way to the physics of the problem, or it might damp out the required solution as well as the spurious terms. The effect of artificial dissipation in equation (17) is shown in Figure 3.

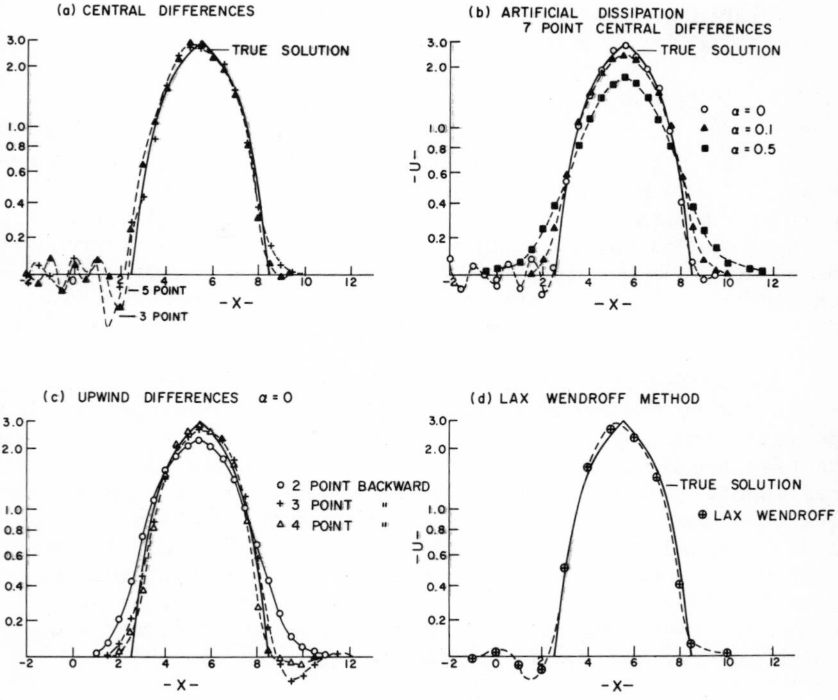

Fig. 3. Analytical Techniques in the First Order Hyperbolic Equation. $\partial u/\partial t = -c\,\partial u/\partial x$

UPWIND DIFFERENCING

First order hyperbolic equations describe a directionally dominated situation. In simple cases, such as equation (15), the direction of propagation is obvious, and the fluid down wind has no premonition of the approaching disturbance. Backward differencing, therefore, gives a physically realistic solution. Schiesser automatically selects first order upwind differencing for hyperbolic equations in the LEANS program(1). The objection to simple backward differences has long been that their first order accuracy is inadequate for most applications, so the Lax Wendroff formulation, which is second order, has been preferred in finite difference approaches(21). However, there is no need to restrict upwind differencing to first order formulae, and higher orders give much improved results as shown in Figure 3.

BOUNDARY CONDITIONS

Boundary conditions for PDE's of second order or lower can be described by some functional relationship

$$\phi\left(u,t,\frac{\partial u}{\partial x}\right)_{x=0} = 0$$

If only the first power of u and $\partial u/\partial x$ appear in this relationship, then the boundary condition is said to be linear, and explicit expressions in terms of neighbouring points may be found for u at the boundary, otherwise an iterative scheme is required. For generality, boundary conditions may always be regarded as non-linear, but it is more efficient to permit the user to specify boundary conditions which are linear, or even constant.

The application of boundary conditions is normally handled internally by the program, but the user is responsible for their specification, and it is at this point again that the claim that the user needs no knowledge of numerical analysis is severely challenged. The equation or set of equations must be given the correct number of boundary conditions, or the problem will be ill-determined and the results spurious. In general, for NE equations of maximum spatial order NX, and temporal order NT, NE*NX boundary conditions and NE*NT initial conditions are required. In the absence of an imposed requirement for a particular boundary, the PDE is assumed to apply at that point.

In systems of hyperbolic equations the problem does not end with determining how many boundary conditions are required, as one must also know at which boundary to apply the relationships. One can envisate a problem in which the number of required boundary conditions changes during the solution, for example, when flow at the boundary reverses in direction. One method of automatically monitoring such changes in hyperbolic systems is to integrate the equations at the boundary by a simple method of characteristics step, thus utilizing the natural properties of the system[20]. The number of characteristics curves crossing the boundary indicate not only the number of required boundary relationships but also provide the appropriate equations[23].

CONSERVATION EQUATIONS

The most frequently encountered set of first order hyperbolic equations are the conservation equations of one-dimensional flow of an ideal gas, which may be written

$$\frac{\partial \bar{u}}{\partial t} + \frac{\partial f(\bar{u})}{\partial x} = 0 \tag{18}$$

where

$$\bar{u} = \begin{bmatrix} \rho \\ e \\ m \end{bmatrix} \qquad f(\bar{u}) = \begin{bmatrix} m \\ m^2/\rho + p \\ (e+p)\ m/\rho \end{bmatrix}$$

and

ρ = mass per unit volume

v = velocity in the x direction

m = momentum per unit volume (ρv)

γ = ratio of specific heats

e = energy per unit volume = $\frac{1}{2}\rho v^2 + P/\gamma\text{-}1$

p = pressure

As in equation (15), there are no dissipative terms, so a disturbance may propagate through the system without dispersing, thus finite discontinuities can develop during the course of the solution. In compressible flow, these discontinuities are called shocks. Standard procedures frequently break down in the presence of shocks as the discontinuities cause infinite derivatives. Once again this can be avoided by adding artificial dissipation term to the equations as in (17) above, but the magnitude of α is hard to assess. Von Neumann and Richtmeyer[22] introduced a pseudo-viscous pressure q into the conservation equations and a rationale to determine the size of this dissipative term.

Recent investigations of the two-phase flow conservation equations recommend the combination of pseudo viscosity and second-order upwind differencing[23-25]. It should be pointed out that in systems of equations, upwind differencing is confined to terms of the type $u(\partial\rho/\partial x)$, backward differencing for $u > 0$, forward for $u < 0$.

The conservation equations in single phase flow will be used to illustrate the simplicity of obtaining adequate solutions with the method of lines approach and the above methods.

Stated in terms of velocity density and pressure, (18) becomes

$$\frac{\partial\bar{u}}{\partial t} = \bar{A}\,\frac{\partial\bar{u}}{\partial x}$$

where

$$\bar{A} = \begin{bmatrix} v & \rho & 0 \\ 0 & v & 1/\rho \\ 0 & P & v \end{bmatrix} \qquad \bar{u} = \begin{bmatrix} \rho \\ v \\ P \end{bmatrix} \tag{19}$$

Dissipation may be included in the equations in the form of friction and pseudo viscosity. The equations then become

$$\begin{aligned} \frac{\partial\rho}{\partial t} &= -v\,\frac{\partial\rho}{\partial x} - \rho\,\frac{\partial v}{\partial x} \\ \frac{\partial v}{\partial t} &= -v\,\frac{\partial v}{\partial x} - \frac{g}{\rho}\left[\frac{\partial (P+q)}{\partial x} + f\right] \\ \frac{\partial P}{\partial t} &= -v\left[\frac{\partial (P+q)}{\partial x} + f\right] - \gamma P\,\frac{\partial u}{\partial x} \end{aligned} \tag{20}$$

where

$f \;=\; \dfrac{\alpha\Delta x\rho v|v|}{2g}$, the frictional resistance

$q \;=\; \dfrac{\beta\Delta x^2\rho}{g}\left(\dfrac{\partial v}{\partial x}\right)^2$, the pseudo viscous pressure, 0 if $\dfrac{\partial v}{\partial x} > 0$

g = gravitational constant

α, β = assignable multipliers

Equations (20) were programmed for solution in FORSIM, and run on the familiar problem of sudden decomposition of a uniform pipe, which is discussed in some detail by Rudinger[26] and Von Rosenberg[27]. Although an apparently simple problem, it effectively illustrates several problems inherent in solving the continuity equations. Figure 4 shows a listing of the routine required to investigate the solution of equations (20), and the results are shown in Figures 5 to 7.

```
C
      SUBROUTINE UPDATE
C
C     COMMUNICATION COMMON BLOCKS
C
      COMMON/INTEGLT/U(51),R(51),P(51)
     ,       /DERIVT/UT(51),RT(51),PT(51)
     ,       /DERIVX/UX(51),RX(51),PX(51)
     ,        /RESERVD/T,DT
     ,       /CONTROL/IOUT
     ,        /PARTS/NCUP,NPOINT,XL,XU,AUX,AUXX,NEQDX,DX,X(1)
     ,       /PARTB/BL(3,10),BU(3,10),NL(10),NU(10)
     ,       /PARAMS/RATIO,ALF,BET
      REAL KG,C(51),Q(51),QX(51),F(51)
      LOGICAL AUXX
      DATA RG,TT,PA,PSF,GA/48.3,500.,14.7,144.,1.4/
      DATA GC/32.2/
C
C        INITIAL CONDITION SECTION
C
      IF(T )100,100,200
 100  AUXX=.F.
      NMAX=51
      NL(2)=NL(3)=-2 $ NL(1)=0
      PR=PA*PSF
      DO 120 I=1,NPOINT
        U(I)=0.
        Q(I)=0.
        P(I)=PR*RATIO
 120    R(I)=P(I)/(RG*TT)
      RR=PR/(RG*TT)
      BU(3,2)=RR
      BU(3,3)=PR
C
C     DYNAMIC SECTION          FIRST ESTABLISH BC.S FOR INFLOW AND OUTFLOW
C
 200  NU(2)=NU(3)=0 $ NU(1)=-2
      DP=PR-P(NPOINT)
      IF(U(NPOINT).GE.0)GOTO220
        BU(3,1)=-SQRT(AMAX1(0.,2*GC*DP/RR))
        NU(1)=0 $ NU(3)=-2
C
C     CALL FORSIM ROUTINES TO APPLY BC.S AND DISCRETIZE DERIVATIVES
C     DEFINE EQUATIONS
C
 220  CALL PARSET(3,NMAX  ,U,UT,UX,UXX)
C
      DO 240 I=1,NPOINT
        Q(I)=-0.5*BET*DX*DX*UX(I)*ABS(UX(I))*R(I)/GC
        IF(UX(I).GT.0)Q(I)=0.
 240    F(I)=0.5*ALF*DX*U(I)*ABS(U(I))*R(1)/GC
      CALL PUPX(Q,QX,NCUP)
C
      DO 260 I=1,NPOINT
        UT(I)=-U(I)*UX(I)-GC*(PX(I)+QX(I)+F(I))/R(I)
        RT(I)=-R(I)*UX(I)-U(I)*RX(I)
        PT(I)=-P(I)*GA*UX(I)-U(I)*(PX(1)+QX(1)+F(I))
 260    C(I)=SQRT(ABS(GA*GC*P(I)/R(I)))
C
      CALL PARFIN(3,NMAX  ,U,UT,UX,UXX)
C
C          PRINTOUT SECTION
C
      IF(IOUT.EQ.0)RETURN
      CALL RITER(X,10HCOORDINATE)
      CALL RITER(U,2H U)
      CALL RITER(R,2H R)
      CALL RITER(P ,2H P)
      CALL RITER(C ,2HC )
      CALL RITER(Q ,2HQ )
      CALL RITER(F ,2HF )
C
      END
```

Fig. 4. Routine for Solution of Conservation Equations of Compressible Flow.

At the instant that the pressurized duct is opened to atmosphere, discharge takes place, and a rarefaction wave travels towards the closed end where it is reflected. As pressure falls below ambient, flow at the open end reverses, and when the wave returns to the open end, it is reflected as a compression wave which becomes a shock. If one assumes no losses, reflection continues indefinitely.

A rarefaction wave does not generate a severe discontinuity and is, therefore, easier to simulate, as shown in Figure 5. The steep front of the shock wave, however, does cause problems. In Figure 6, for $\beta=0$ with three-point central differences used throughout, the solution is subject to spurious waves which compound on reflection, eventually causing infinite derivatives in the shock. For $\beta=4$, the spurious oscillations are considerably reduced, while $\beta=9$ permits the solution to continue free of numerical instability. Realistic values of the friction parameter α do not significantly affect the solution, but Von Neumann and Richtmeyer show that values of β up to 4g are permissible.

Using upwind differencing also reduces the spurious oscillations, as shown in Figure 7, and the best solution is obtained by using a combination of pseudo viscosity and upwind differencing. All these possible modes of solution may be easily implemented within the framework of the method of lines approach.

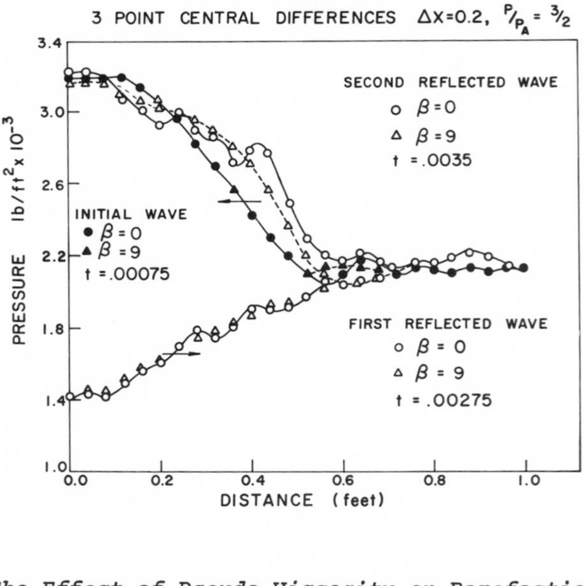

Fig. 5. The Effect of Pseudo Viscosity on Rarefaction Wave Analysis.

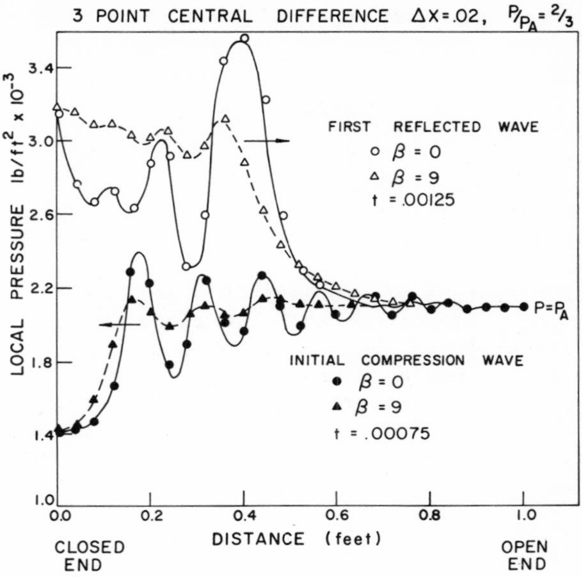

Fig. 6. The Effect of Pseudo Viscosity in Shock Wave Analysis.

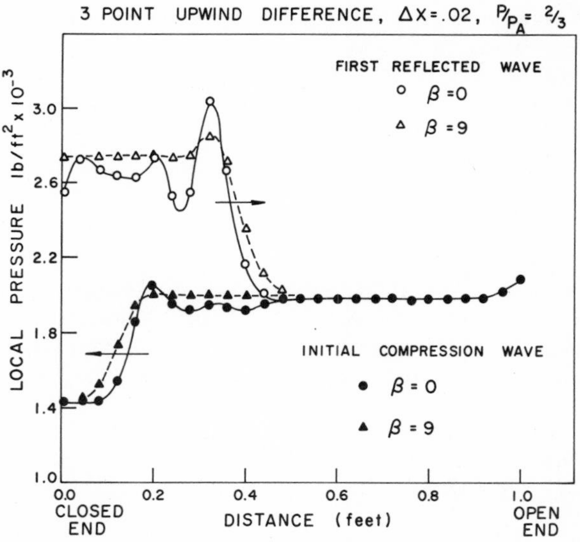

Fig. 7. Upwind Differencing in Shock Wave Analysis.

INCREASING EFFICIENCY FOR PRODUCTION RUNS

Apart from using the optimizing compiler and following the principles of good programming practice within his own routines, the user has the opportunity to streamline the flow within the FORSIM system routines. Derivatives of each variable have individual control, so unwanted derivatives need not be evaluated. For simple boundary conditions, the user may apply his own within the UPDATE routine and bypass the general boundary routine.

For large systems of coupled ODE's and PDE's, considerable improvements of performance can be obtained by separating the equations into stiff and non-stiff subsets. The stiff set, usually the PDE's, may then be integrated with Gear's algorithm while the remaining equations are integrated by less sophisticated means, as their accuracy and stability are less critical. This division of responsibilities has been extremely successful in solving reactor kinetics problems[10].

CONCLUSIONS

Obviously there are pitfalls in using an automated package for the solution of PDE's, but provided the user is aware of the possibilities of behaviour of his equations, many problems which would otherwise require detailed programming effort may be solved in a straight-forward manner. Results from the method of lines approach are closely reproducible even when different component algorithms are used, but probably the greatest advantage offered by the method is its versatility, which permits extra terms to be added to the equations without increasing the complexity of the programming required.

REFERENCES

(1) Schiesser, W.E., A Digital Simulation System for Mixed Ordinary/Partial Differential Equation Models, Proceedings, IFAC Symposium on Digital Simulation of Continuous Systems, Gyor, Hungary, 2, p.S2-1 to S2-9, September 1971.

(2) Larsen, L.A., Automatic Solutions of Partial Differential Equations, University of Illinois at Urbana-Champaign Report UIUCDCS-R-72-546, October 1972.

(3) Sincovec, R.F. and Madsen, N.K., Software for Nonlinear Partial Differential Equations, Proceedings, Conference on Mathematical Software II, Purdue University, May 1974.

(4) Loeb, A.M., User's Guide to a New User-Oriented Subroutine for the Automatic Solution of One-Dimensional Partial Differential Equations, Proceedings, 1974 Summer Computer Simulation Conference, Houston, Texas, p.95-103.

(5) Carver, M.B., A FORTRAN Oriented Simulation System for the General Solution of Partial Differential Equations, Proceedings, 1973 Summer Computer Simulation Conference, SCI, La Jolla, p.46-50.

(6) Ahmad, S.Y. and Carver, M.B., A System for the Automated Solution of Sets of Implicitly Coupled Partial Differential Equations, Proceedings, 1974 Summer Computer Simulation Conference, Houston, Texas, p.104-112.

(7) Carver, M.B. and Baudouin, A.P., Solution of Reactor Kinetics Problems Using Sparse Matrix Techniques in an ODE Integrator for Stiff Equations, Atomic Energy of Canada Limited Report AECL-5177, January 1976.

(8) Gear, C.W., The Automatic Integration of Ordinary Differential Equations, *Comm. ACM*, 14, 3, p.176-180, 185-190, March 1971.

(9) Hindmarsh, A.C., GEAR: Ordinary Differential Equation System Solver, Lawrence Livermore Laboratory Report UCID 300001, Rev. 2, August 1972; and GEARB: Solution of Ordinary Differential Equations Having Banded Jacobian, Lawrence Livermore Laboratory Report UCID 30059, May 1973, Rev. 1, March 1975.

(10) Baudouin, A.P. and Carver, M.B., The Solution of 1D Reactor Kinetics Problems by the Method of Lines Combined with an ODE Integrator for Stiff Equations, Advances in Computer Methods for Partial Differential Equations, p.377-380, R. Vichnevetsky, Editor, AICA Press 1975, Rutgers University, New Jersey.

(11) Curtis, A.B. and Reid, J.K., FORTRAN Routines for the Solution of Sparse Sets of Linear Equations, AERE-R-6844, Harwell, 1971.

(12) Carver, M.B., FORSIM: A FORTRAN Package for the Automated Solution of Coupled Partial and/or Ordinary Differential Equation Systems User's Manual, Atomic Energy of Canada Limited Report AECL-4844, November 1974.

(13) Collatz, L., Hermitian Methods for Initial Value Problems in Partial Differential Equations, Conference on Numerical Analysis, Riana, Dublin, Ireland, 1972.

(14) Bellman, R. and Kashef, B.G., Application of Spline and Differential Quadrature to Partial Differential Equations, 59th Hawaii International Conference on Systems Science, Honolulu, 1972, also Differential Quadrature: A Technique for the Tapid Solution of PDE's, *J. of Computational Physics, 10,* p.40-52, 1972.

(15) Finlayson, B.A., The Method of Weighted Residuals and Variational Principles, Academic Press, New York, 1972.

(16) Vichnevetsky, R. and Peiffer, B., Error Waves in Finite Difference and Finite Element Methods for Hyperbolic Equations, Advances in Computer Methods for Partial Differential Equations, p.53-59, AICA Press, Rutgers University, 1975.

(17) Murphy, W.D., Cubic Spline Galerkin Approximations to Parabolic Systems with Coupled Nonlinear Boundary Conditions, *Int. J. for Numerical Methods in Engineering, 9,* p.63-71, 1975.

(18) Loeb, A.M. and Schiesser, W.E., Stiffness and Accuracy in the Method of Lines Solution of Partial Differential Equations, Proceedings, 1973 Summer Computer Simulation Conference, SCI, La Jolla, p.25-39.

(19) Jones, D.J. et al., On the Numerical Solution of Elliptic Partial Differential Equations by the Method of Lines, *J. Comp. Phys. 9,* p.496-527, June 1972.

(20) Vichnevetsky, R. and Tomalesky, A.W., Spurious Wave Phenomena in Numerical Approximations of Hyperbolic Equations, Proceedings, Fifth Annual Princeton Conference on Information Science and Systems, p.357-363, March 1971.

(21) Richtmeyer, R.D. and Morton, K.W., Difference Methods for Initial-Value Problems, Interscience Publishers, New York, 1957.

(22) Von Neumann, J. and Richtmeyer, R.D., A Method for the Numerical Calculation of Hydrodynamic Shocks, *J. Appl. Phys., 21,* p.232, 1950.

(23) Mathers, W.G., Method of Lines Solution of the Homogeneous Equilibrium Flow Boiling Problem, Atomic Energy of Canada Limited, Whiteshell Nuclear Research Establishment, private communication, March 1974.

(24) Patankar, S.V. and Spalding, D.B., A Calculation Procedure for Heat, Mass and Momentum Transfer in Three-Dimension Parabolic Flows, *Int. J. Heat Mass Transfer,* *15*, p.1787-1806.

(25) Ahmad, S.Y. and Carver, M.B., Automatic Treatment of Shocks in Two-Phase Flow, Proceedings, 5th Canadian Congress of Applied Mechanics, p.571-572, Fredericton, May 1975.

(26) Rudinger, G., Wave Diagrams for Non-Steady Flow in Ducts, Van Nostrand, 1955.

(27) Von Rosenberg, D.U., Beauchamp, D.L., and Watts, J.W., An Efficient Numerical Solution of Non-Linear Hypberbolic Equations with Split Boundary Conditions, *Chemical Engineering Science, 23,* p.345-351, 1968.

(28) Elliot, L.A., Computing Experience with Hyperbolic PDE's, *Proc. Roy. Soc., London, A, 323,* p.263-270, 1971.

Panel Discussion of Quality Software for ODEs

G. D. Byrne, C. W. Gear, A. C. Hindmarsh, T. E. Hull,
F. T. Krogh, and L. F. Shampine

1. Introduction (G. D. Byrne).

This paper contains the formal statements made by the panelists in the order of their presentations at the 80th National A.I.Ch.E. meeting, as well as a summary of the question and answer session which followed the formal statements. The purpose of the panel discussion was to provide an exchange of information between developers and designers of ordinary differential equation (ODE) software on one hand and a group of users or potential users of the software on the other. Consequently, the topics which the panelists presented included: general attributes and availability of ODE software, the necessity of code modifications, the ODE codes in use at Lawrence Livermore Laboratory, the measurement of reliability and efficiency of software by testing, test results obtained by varying strategies within a particular code, and properties of two ODE packages in use at Sandia Laboratories.

2. Some Features of Quality Codes for ODE's* (G. D. Byrne)

The software for solving ordinary differential equations (ODE's) that concerns us here involves much more than a simple implementation of a formula for solving an ODE (or a system of

---

*This work was performed under the auspices of the USERDA, while the author was on leave at the Applied Mathematics Division, Argonne National Laboratory, Argonne, IL 60439.

ODE's) with a prescribed initial value. Each of these codes makes an error estimate and adjusts the step size (or time step) in an effort to keep the error below a value related to a tolerance supplied by the user. The ODE solvers using methods of backward differention type (e.g., DIFSUB, GEAR, EPISODE, etc.), of Adams type (e.g., DIFSUB, DVDQ, GEAR, DE/STEP, INTRP, EPISODE, etc.), or other linear multistep methods also change orders (switch formulas) automatically. Most of these codes also provide a driver subroutine which might perform such chores as checking the user's input parameters for errors, writing error messages, setting error flags, checking to see if the problem can be solved by the particular integrator called, repeatedly calling the step-by-step integrator, performing interpolation to provide a numerical solution at a prescribed point, and so on. Thus, the driver relieves the average user of much responsibility and makes the ODE solver convenient for him to use.

The typical code is written in readable FORTRAN and, with minor modifications, can be used on several target machines. The documentation consists of the comments in the code and supporting publications such as journal articles, technical reports, and books. The documentation explains the underlying algorithm, the use of the code, and frequently possible modifications for specific applications. Each ODE solver also reflects the environment in which it was designed and implemented and, hence, solves some class of problems well.

At the present, several of these codes can be obtained free of charge by sending a magnetic tape, tape specifications, machine description, and name and number of the code sought to Argonne Code Center, Applied Mathematics Division, Argonne National Laboratory, 9700 South Cass Avenue, Argonne, IL 60439 (telephone 312-739-7711, ext. 4366). These ODE solvers include: GEAR, Argonne Code Center (ACC) Abstract #592, written by A. C. Hindmarsh; DE/STEP, INTRP, ACC Abstract #640, written by L. F.

Shampine and M. K. Gordon; GEARB, ACC Abstract #661, written by A. C. Hindmarsh; and EPISODE, ACC Abstract #675, written by G. D. Byrne and A. C. Hindmarsh. Upon its return, the magnetic tape will be accompanied by a user's guide for each requested code. Other ODE solvers may be obtained directly or indirectly through the author-panelists.

3. The Use of Quality O.D.E. Codes in User Packages (C. W. Gear).

The following remarks are not restricted to O.D.E. codes, but such codes represent one of a few areas for which codes and methods are sufficiently complex to present many problems not found in simple codes such as SIN routines, but for which methods are well enough understood that general purpose quality codes have been written. These codes are typically used in two ways: (a) directly -- the user provides the required derivatives and execution control statements (e.g. by subroutine calls in Fortran), and (b) indirectly -- the code is part of a much larger user oriented package (e.g. a package which accepts a statement of chemical reaction equations, converts these to derivative evaluation code, and calls the integrator for the user). In the first mode of use, the code can be obtained by the user from the library, and the library version can be a direct copy of a carefully written and checked, well documented code prepared by a person who has invested a lot of time in studying the algorithms used and their machine implementation. Those are necessary (but not sufficient) conditions for quality software. In O.D.E.'s, such codes are available from my fellow panelists Hindmarsh, Krogh, and Shampine or the Argonne Code Center. In the second mode of use, the code must frequently be modified to interface to the user package and its host machine(s). (The latter modifications are also needed in library code distributed across machine boundaries.) These modifications are necessary because the code forms only a small part of the total package, and it must meet the demands imposed by requirements of efficiency and flexibility of the total package. If the code were organized as a subroutine with many options, it would be hopelessly inefficient for a subclass of problems, whereas if it lacked flexibility it would still have to be modified to get the desired characteristics.

Of necessity, the modifications must be done by people whose primary expertise is in other areas, and this introduces several problems. Firstly, modified quality software seldom retains its quality, particularly if it is numerical software. Secondly, even quality software will be found to contain errors, and the underlying methods will be improved as theoretical understanding improves. Changes to correct errors and implement improvements can be distributed to computer centers and find their way into the library. It is extremely difficult for such changes to be incorporated into versions of the code that have been modified. The result is that most of the numerical sections of large user oriented packages directed at the community of non-programmers consist of questionable implementations of out-of-date methods.

The solution has to lie in schemes for the distribution of quality software that has been designed to be modified. The designer must have considered which modifications are acceptable, and which are not. If the package programmer is fortunate enough to work with Hindmarsh, Krogh, or Shampine, he can obtain a version of their integrators suitably modified. However, in the majority of cases, the code designer must prepare a code which clearly indicates which sections are essential and cannot be modified, and which sections can be tailored to the needs of the package. This distinction may have to specify which blocks of code must be inviolate and which can be changed, it may have to specify which sections of a single statement are essential. For example, some sections of initialization code could be moved from the subroutine to the main program and changed whereas other sections must retain their relationship to the rest of the code, or within a single statement the order of calculation may be important whereas the way in which data is stored as lists or vectors may not be important.

If future changes in codes are to be incorporated into existing packages, the modifications must be done mechanically -- that is, the code must be written in a high level language that can be edited at a high level. A mechanism for this is the preprocessor or macro-translator. The essentials of the code should be coded, while the features needed by the package designer (organization of storage, etc.) provided as a series of options or left for later specification. Preprocessing of the code can remove the unwanted options and include those specified by the package programmer. If "system generation" of the package includes this preprocessing of the library version of the code using a set of specifications provided by the package programmer, new versions of the code can be incorporated by another system generation.

This mode of operation is not with us yet, and the emphasis of those of us who write software for distribution has been on flexible subroutines with some thought for the problem of the changes needed for different word lengths and arithmetic. My experience with one code for O.D.E.'s has been that it has been extensively changed -- the typical comment has been that it does not provide the particular option needed, and that there are too many parameters for the specification of options not needed! I would like to be able to distribute a code that would meet most requirements -- but for this we need some developement of the sort I have outlined.

4. Quality Software for Ordinary Differential Equations at LLL* (A. C. Hindmarsh).

Work at the Lawrence Livermore Laboratory in the area of ordinary differential equations (initial value problem) has been motivated by a variety of problem areas. One of these is the study of chemical kinetics systems, and another is the treatment of systems of partial differential equations (such as arise in atmospheric models) by the method of lines. In both of these, the ODE problem is typically stiff, and in the second area, the size of the problem is typically quite large as well. Thus, much of the quality ODE software in use at LLL has been developed for stiff systems, with emphasis on extendability to large stiff systems, although not to the exclusion of quality software for nonstiff problems.

The basic element in the LLL array of ODE software is a general purpose package called GEAR [1]. This is a package of seven subroutines (used together as a whole) which is based on C. W. Gear's well known subroutine DIFSUB, but which involved a complete rewrite of that routine, with significant departures in both the basic methods and the user interface. For example, the code includes eight different method options, in that the choice of basic formula (Adams or backward differentiation) is independent of the choice of corrector iteration method (of which there are four). Also, for example, the package includes a driver which is called by the user and in turn calls other routines until the desired output points are reached. Most of the internal arrays are stored in Common, as this saves on indirect addressing, with observed run time reductions of up to 13%.

A second basic code, called EPISODE, was developed more recently by G. D. Byrne and myself [2] and reported on at this meeting. To the user, it appears very similar to GEAR, but

---

*Work performed under the auspices of the U.S. Energy Research and Development Administration. Contract Number W-7405-ENG-48.

differs internally in some radical ways. The most important is the use of truly variable-step formulas throughout. This is somewhat more expensive to implement, but it has the effect of eliminating a potential instability of GEAR that tends to arise when the ODEs have rather wildly behaving driving terms (for example, in chemical kinetics with diurnally varying rates). Thus, the EPISODE package is a valuable supplement to, but not a replacement for, the GEAR package.

As general ODE packages, GEAR and EPISODE are quite useful for both stiff and nonstiff problems. In the nonstiff case, with the nonstiff method option, they seem to perform competitively in comparison with other sophisticated nonstiff system solvers, such as the DE/STEP package of Shampine and Gordon [3]. In the stiff case, these codes allow for (and in fact require, for the sake of efficiency) the use of the Jacobian matrix, and contain routines for solving the associated linear systems, in full matrix form. These matrix computations are minimized by using efficient factorization and backsolution routines [4] and by retaining the factored matrix over many steps before reevaluating the Jacobian.

In many applications, however, the use of GEAR or EPISODE is prohibited simply by the large size of the problem, which, because of stiffness, would necessitate the use of a matrix too large to store. As a result, several variants of these packages have been written in which more specialized matrix routines are substituted. Thus, band matrix variants GEARB [5] and EPISODEB were written for the case of a banded Jacobian. Also, a variant called GEARIB was written to solve implicit systems of the form $A(y,t)\dot{y} = g(y,t)$, with banded treatment of the matrices A and $\partial g/\partial y$. The GEARB and GEARIB packages have been used quite successfully by N. K. Madsen and R. F. Sincovec to construct automatic solvers for one-dimensional partial differential equation systems [6,7]. Two other variants of GEAR have been recently developed -- one for the case of a general sparse Jacobian [8] and one in which vectorized language features have been used [9].

For application to two-dimensional problems, the above variants are of rather limited value (e.g., the bandwidth tends to become prohibitively large). In order to solve the two-dimensional atmospheric models developed at LLL, the matrix specialization that was chosen is block-SOR, in which the size of the block is the number of chemical species, and the number of blocks in each direction along the matrix is the number of spatial zones [10]. The associated variant of the GEAR package was further specialized to use Large Core Memory storage of most of the arrays, on the CDC 7600. To give an example of the capabilities of the latter code, runs were made for the LLL stratospheric kinetics-transport model with 9 species and 1628 zones (14652 ODE's), over a time span of 10^9 sec, and with an initial time step (forced by the stiffness of the kinetics) of 10^{-10} sec.

REFERENCES

[1] A. C. Hindmarsh, GEAR: Ordinary Differential Equation System Solver, Lawrence Livermore Laboratory Rept. UCID-30001, Rev. 3, 1974.

[2] G. D. Byrne and A. C. Hindmarsh, A Polyalgorithm for the Numerical Solution of Ordinary Differential Equations, ACM Trans. Math. Software 1, (1975), pp. 71-96.

[3] L. F. Shampine and M. K. Gordon, Computer Solution of Ordinary Differential Equations: The Initial Value Problem, W. H. Freeman and Co., San Francisco, 1975.

[4] C. B. Moler, Algorithm 423, Linear Equation Solver, Comm. ACM 15 (1972), p. 274.

[5] A. C. Hindmarsh, GEARB: Solution of Ordinary Differential Equations Having Banded Jacobian, Lawrence Livermore Laboratory Rept. UCID, Rev. 1, 1975.

[6] R. F. Sincovec and N. K. Madsen, Software for Nonlinear Partial Differential Equations, ACM Trans. Math. Software 1 (1975), pp. 232-260.

[7] N. K. Madsen and R. F. Sincovec, Generalized Software for Partial Differential Equations, this volume.

[8] J. W. Spellmann and A. C. Hindmarsh, GEARS: Solution of Ordinary Differential Equations Having a Sparse Jacobian Matrix, Lawrence Livermore Laboratory Rept. UCID-30116, 1975.

[9] D. B. Morris and A. C. Hindmarsh, GEARV: A Vectorized Ordinary Differential Equation Solver, Lawrence Livermore Laboratory Rept. UCID-30119, 1975.

[10] J. S. Chang, A. C. Hindmarsh, and N. K. Madsen, Simulation of Chemical Kinetics Transport in the Stratosphere, in Stiff Differential Systems, R. A. Willoughby, Ed., Plenum Press, New York, 1974, pp. 51-65.

5. Testing and Certification of ODE Programs (T. E. Hull).

The design of a computer program must of course be determined mainly by the needs of the users. The environment in which the program is to be used will also have an important influence. To a large extent these requirements will motivate the development of library programs that are easy to use, and that are easily adapted to specific needs.

Within the constraints imposed by these requirements, there are two basically different properties that good programs ought to have. One is reliability; the other is efficiency. It is the purpose of this talk to comment very briefly on each of these topics, with special emphasis on the importance of careful testing. Some specific examples will be used to illustrate.

With regard to reliability, the main purpose in testing a program is to help determine the correctness of the program, in the sense of helping make sure that the program accomplishes what is claimed of it. This involves testing the accuracy of the program, trying to test each of the different options provided by the program, and attempting to delineate the domain over which it behaves properly (including the likelihood that it will fail-safe, in some sense, if it is ever used outside its domain of validity).

Of course, no amount of testing will settle these questions completely, and it is important to support claims about correctness with as much theoretical evidence as possible. Proofs about certain properties are possible, particularly about those depending on combinatorial properties of the program. Proofs about accuracy are possible too, but only over relatively small parts of its domain. However, even these limited theoretical results, in combination with careful testing to determine the limits to its domain, are highly desirable components in the certification of a program.

A second purpose in testing a program is to help provide measures of its efficiency. It is important here to make a clear distinction between measuring the efficiency of a particular program, and making comparisons between different programs. Two different programs are rarely designed to accomplish the same tests, so direct comparisons are not usually possible. Direct comparisons become possible, under the same set of test conditions, only in special circumstances. An example would be the comparison of different formulas to be used in the same context. This sort of comparison is helpful in choosing the components of a program that is to be developed into a useful piece of numerical software. But trying to compare existing codes can lead one to make changes in some of the codes, so that the resulting comparisons are actually valid only for the modified codes. Although some useful conclusions can still be drawn from the resulting comparisons, it will be necessary to test the unmodified codes as well. Theoretical work on efficiency is helpful in understanding general concepts such as stability, rates of convergence, and so on. But a good theoretical treatment of efficiency, in the spirit of computational complexity, is only just beginning to be a possibility.

As already stated, user needs will impose essential constraints on the design of ODE programs. However, within these requirements, it is important to standardize as much as possible in order to facilitate certification procedures. A standard calling sequence would contribute to ease of use, and enable the user to make more meaningful comparisons, especially if standard criteria are also required of different programs. A good standard structure for the program itself would also facilitate comparisons, as well as making it much easier to prove properties of the different programs.

6. Summary of Test Results with Variants of a Variable Order Adams Method* (F. T. Krogh).

A variable order Adams method can be implemented and/or used in many different ways. In the process of developing a variable order Adams code which uses modified divided differences for changing stepsize, see [1] or [2], we have had occasion to explore the effect of four parameters on the performance of the program. The effect of these parameters was studied for test problems 8 and 9 of [3]. Both are simple two-body problems, one with circular motion, and the other traces out an ellipse with eccentricity .6. For each of these problems we have examined all possible combinations of two distinct possibilities for each of the parameters at requested error tolerances of 10, 1, .1, ..., 10^{-20} ($2\times16\times22$ cases). The choices examined are given below.

I. Use a PECE or a PEC Adams method, always with a corrector that has order one greater than the predictor. (It is easy to show for such methods that a PEC method is equivalent to a PE method, where the P is the same order as the C in the PEC method.)

II. Use either (2, 1/2) or (9/8, 7/8) as nominal factors for changing the stepsize. The program allows other values, but these values are probably extreme cases. The program does not restrict itself to the nominal factors when

*This paper presents the results of one phase of research carried out at the Jet Propulsion Laboratory, California Institute of Technology, under Contract NAS7-100, sponsored by the National Aeronautics and Space Administration.

TWO-BODY PROBLEM, CIRCULAR MOTION,
WITH AN ABSOLUTE ERROR TEST
RESULTS ON THE INTERVAL (0, 16 PI)

LOG OF ERROR TOLERANCE:				-6		-9		-12	
ND/ STEP	HINC	EQ. ORD.	INT. ORD.	ABSOLUTE ERROR	ND	ABSOLUTE ERROR	ND	ABSOLUTE ERROR	ND
2	2	2	SAME	(-5)2.0	299	(-8)2.0	409	(-12)6.7	626
2	2	2	DIFF	(-5)2.7	303	(-9)3.6	455	(-13)4.8	682
2	2	1	SAME	(-4)2.5	491	(-8)2.5	811	(-10)1.0	1347
2	2	1	DIFF	(-4)3.1	493	(-8)5.3	811	(-10)2.4	1229
2	9/8	2	SAME	(-5)4.3	279	(-8)2.0	409	(-12)8.0	610
2	9/8	2	DIFF	(-5)3.3	303	(-8)2.1	409	(-12)8.6	612
2	9/8	1	SAME	(-4)4.0	461	(-8)2.5	811	(-10)2.7	1235
2	9/8	1	DIFF	(-4)3.5	485	(-8)4.8	779	(-10)1.6	1258
1	2	2	SAME	(-6)9.0	269	(-8)3.6	410	(-12)2.7	666
1	2	2	DIFF	(-6)1.5	301	(-8)2.3	404	(-12)9.3	612
1	2	1	SAME	(-4)2.0	777	(-7)2.6	1476	(-10)4.4	2515
1	2	1	DIFF	(-4)1.8	743	(-8)6.9	1422	(-10)2.2	2503
1	9/8	2	SAME	(-5)1.3	252	(-8)3.2	413	(-12)2.0	619
1	9/8	2	DIFF	(-5)2.2	248	(-8)2.9	390	(-12)9.4	607
1	9/8	1	SAME	(-5)8.0	807	(-7)1.8	1505	(-9)1.5	2488
1	9/8	1	DIFF	(-4)1.6	745	(-8)3.3	1441	(-10)2.9	2440
DVDQ				(-8)4.0	411	(-9)2.0	469	(-14)4.5	857
DVDQ(@-5,-8,-11)				(-4)3.4	317	(-9)2.2	442	(-10)1.0	726
STEP				(-4)1.6	494	(-8)6.6	974	(-9)1.3	1225

TWO-BODY PROBLEM, ECCENTRICITY= 6,
WITH AN ABSOLUTE ERROR TEST
RESULTS ON THE INTERVAL (0, 16 PI)

LOG OF ERROR TOLERANCE:				-6		-9		-12	
ND/ STEP	HINC	EQ. ORD.	INT. ORD.	ABSOLUTE ERROR	ND	ABSOLUTE ERROR	ND	ABSOLUTE ERROR	ND
2	2	2	SAME	(-3)1.6	1557	(-7)8.5	2655	(-10)5.2	3935
2	2	2	DIFF	(-4)2.9	1637	(-7)5.1	2698	(-10)3.4	3909
2	2	1	SAME	(-3)3.1	1602	(-7)9.3	2776	(-10)7.7	4623
2	2	1	DIFF	(-3)1.4	1670	(-6)1.0	2768	(-9)1.2	4388
2	9/8	2	SAME	(-3)2.7	1413	(-6)1.6	2150	(-10)3.5	3393
2	9/8	2	DIFF	(-3)1.8	1456	(-6)1.2	2195	(-10)4.9	3401
2	9/8	1	SAME	(-3)3.3	1416	(-6)2.4	2372	(-9)1.4	3817
2	9/8	1	DIFF	(-3)2.4	1500	(-6)1.5	2415	(-10)7.7	3730
1	2	2	SAME	(-4)9.5	819	(-7)1.9	1379	(-10)2.1	2127
1	2	2	DIFF	(-4)2.2	843	(-7)4.6	1425	(-10)3.5	2111
1	2	1	SAME	(-4)4.9	1318	(-7)9.5	2664	(-10)2.3	4865
1	2	1	DIFF	(-3)1.9	1221	(-6)1.6	2729	(-9)2.0	4716
1	9/8	2	SAME	(-3)1.1	722	(-7)4.7	1177	(-11)5.5	1820
1	9/8	2	DIFF	(-4)2.0	732	(-7)6.6	1209	(-10)1.5	1828
1	9/8	1	SAME	(-4)5.9	1284	(-6)1.3	2661	(-10)3.3	4935
1	9/8	1	DIFF	(-4)8.6	1179	(-6)1.8	2426	(-9)1.6	4435
DVDQ				(-4)2.3	2021	(-7)2.8	3131	(-10)6.3	4306
STEP(@-7,-10,-13)				(-3)1.2	1754	(-7)4.9	3017	(-10)6.6	5666

changing the stepsize, but factors closer to one do tend to cause more frequent changes in stepsize, more integration overhead, and fewer function evaluations.

III. Solve the 2nd order equations directly, or solve them by breaking them up into an equivalent system of first order equations.

IV. Use the same integration orders for all equations or select the integration orders independently for each equation.

The results given for DVDQ on these problems in [3] were obtained using a PECE method, (2, 1/2) as the <u>only</u> factors for changing the stepsize, solving the 2nd order equations directly, and selecting integration orders separately for different equations. In its current state, and with options as close to DVDQ as possible, the integrator used for the comparisons here was of approximately the same efficiency for the case of circular motion (although the global error was a more regular function of the local error tolerance as a result of not restricting changes in the stepsize to the nominal factors), and approximately 10-20% more efficient in the case of elliptic motion (which we attribute primarily to the different procedure for changing stepsize). Performance of the new integrator would have been somewhat better if we had allowed it as large a maximum integration order as DVDQ. We plan to investigate this (and some other things) in more detail at a later time.

We summarize below the results of our tests. Generalizations to other problems are risky of course.

1. There was no appreciable difference between using the same and using different integration orders. We believe the primary advantage of using different integration orders comes when integrating equations with different characteristics. And we have encountered problems where one is better off requiring the integration orders to be the same on certain equations.

2. Integrating the second order equations directly is always best (of course, problems are known for which this is not true, see e.g. [4]); in the case of circular motion better by a factor of over two, except for the PECE methods at low accuracy where the factor is about 1.5. In the case of the elliptic motion results are somewhat similar, except that for the PECE methods the direct integration offers only a small advantage. We attribute the results to the better stability characteristics of the direct integration of second order equations on these problems. The cases where the differences are not great are those where discretization error is the primary factor limiting the stepsize.
3. It is not as well known as it should be that the primary advantage of doing two derivative evaluations per step is due to improved stability characteristics and not to a smaller error term. Thus on the elliptic problem when integrating 2nd order equations directly, the PEC method is nearly twice as efficient as the PECE method. (The implications of this for practical problems is not highly significant, since in most problems of this type a predict-partial-correct scheme should be used. Such a scheme computes the main part of the derivative twice per step, and perturbing forces, which typically are significantly more complicated, only after predicting. The values computed after predicting are used in the computation of the derivatives after correcting.) But there are implications for the comparisons of methods, since many methods don't offer the partial correct option. If a comparison is to be made with the best Adams method on such problems, one should compare with a PEC method integrating the second order equation directly. We have obtained similar results on the restricted 3-body problem in [3].

4. Changing stepsize by factors close to one results in a definite reduction (10-20%) in the number of derivative evaluations for the problem which calls for a wide variation in the stepsize. When derivatives evaluations are expensive, factors close to one appear to be a good idea, but when derivative evaluations are cheap, the factors (2, 1/2) seem to be a reasonable choice to reduce the integration overhead. There is not much difference for the case of circular motion, but there is also not much difference in overhead since, except for extreme values of the error tolerance, the integrator tends to use a constant stepsize ultimately. (This would not be true for a problem which is best integrated with slight variation in the stepsize, however.)

REFERENCES

[1] Krogh, Fred T., Changing stepsize in the integration of differential equations using modified divided differences. Proceedings of the Conference on the Numerical Solution of Ordinary Differential Equations, D. G. Bettis, Ed., Lecture Notes in Mathematics No. 362, Springer-Verlag, New York, 1974, pp. 22-71.

[2] Shampine, L. F. and Gordon, M. K., Computer Solution of Ordinary Differential Equations, The Initial Value Problem, W. H. Freeman and Co., San Francisco, 1975.

[3] Krogh, Fred T., On Testing a subroutine for the numerical integration of ordinary differential equations, J. ACM 20 (1973), pp. 545-562.

[4] Krogh, Fred T., A variable step variable order multistep method for the numerical solution of ordinary differential equations, Information Processing 68, North Holland Publishing Co., Amsterdam, 1969, pp. 194-199.

7. Quality Software for Non-Stiff Ordinary Differential Equations (L. F. Shampine).

Codes for solving non-stiff ordinary differential equations have been improved greatly in the lat few years. The major areas of concern are efficiency, ease of use, and safety. By the last we mean the ability to diagnose or otherwise cope with mistakes or misuse of the code. Despite the general progress being made in producing quality software, there is surprising variation in these three areas even among the best codes. This can be seen in a survey made by S. M. Davenport, L. F. Shampine, and H. A. Watts which is to appear in the SIAM Review. Here we wish to indicate the state of software development via two specific codes DE and RKF45 which are members of a systematized collection of codes called DEPAC. According to the survey just mentioned, the Adams code DE is as efficient as any of the codes tested if the equations are expensive to evaluate, and the Runge-Kutta code RKF45 is as efficient as any if the equations are cheap. Granted that these are very efficient codes, how easy to use and how safe are they?

We wish to integrate $y' = f(x,y)$, $y(a)$ given, from $x = a$ to a sequence of points b_1, b_2, ... where we want approximations to $y(b_1)$, $y(b_2)$, ... The two codes have the same call list: F,NEQN,Y,T,TOUT,RELERR,ABSERR,IFLAG. The arguments are very nearly all required to even define the problem. F is the name of a subroutine for evaluating $f(x,y)$. NEQN is the number of equations. When the code is called the first time, Y is the vector of initial values $y(a)$ and T = a, TOUT = b_1. At each internal step the error is to be controlled so that $|\text{local error}| \leq \text{RELERR}* |\text{solution}| + |\text{ABSERR}|$, a mixed relative-absolute error criterion. IFLAG reports if the integration is successful, or if not, why. On return, T is set to the point closest to TOUT where a good solution was computed, so normally T = TOUT. In any event, Y is the solution at T. To go to another output point b_2 after a successful integration, one need only define TOUT = b_2 and call the code again. If the integration is unsuccessful, the user need take only the action specified by IFLAG and call the code again. No other concession has to be made to the efficiency of the code than to call it the first time with IFLAG = 1 so it can initialize. Also with DE if it is impossible to integrate past TOUT, the code should be told this by setting IFLAG = -1.

Input arguments are checked very carefully, RKF45 going so far as to verify that the user has taken the action specified by IFLAG. The codes select their own starting step size, restart as needed, and arrange to produce output at TOUT efficiently. If one requests output too frequently, RKF45 becomes inefficient and tells the user to mend his ways or else to turn to DE which handles this matter very efficeintly. Both codes monitor the number of equation evaluations to provide the user with a cost control. The codes are intended for problems with smooth, moderately stable solutions. Yet they will efficiently and automatically solve problems with integrable singularities and discontinuous first derivatives. Both are relatively efficient when confronted with moderate stiffness; DE diagnoses severe stiffness. Both codes detect requests for impossible accuracies. RKF45 should not be used for very high accuracies, and will not attempt them. DE will take special action to get all the accuracy possible and when detecting an impossible request, inform the user what is possible.

The software design exemplified here will solve most nonstiff problems conveniently. If the user needs more control, there is a way to get this in RKF45 by having the code return after each internal step. The code DE is actually one of a suite of codes, DE/STEP, INTRP, in DEPAC. The associated code STEP is to be used when maximum flexibility is required and is correspondingly somewhat more complicated to use. The present version of DEPAC also contains a unique code called GERK. This code estimates the true, or global, error it makes when solving an initial value problem. It is a Runge-Kutta code, closely related to RKF45, which is surprisingly efficient and reliable when applied to problems with smooth solutions.

8. Question and Answer Session

QUESTION: Do you foresee the development of ODE codes for equations of order greater than two?

ANSWER: (Panel) In chemical engineering, the most important of the orders greater than one is two. Of course the higher order equation can always be reduced to a system of first order ODE's or Fred Krogh's DVDQ can be used to solve higher order problems directly.

QUESTION: We are using spatial discretization and an ODE package to solve a parabolic partial differential equation by the numerical method of lines. The resulting system of stiff differential equations is well-behaved, but the solution experiences a discontinuity or blows up at large values of t. Have you any advice?

ANSWERS: (Gear) Try EPISODE. You might also try setting a switch point short of the discontinuity, integrating just beyond the switch point, and then restarting at the switch point.

(Krogh) My integrators provide for handling this sort of problem via something I call a G-stop.

(Byrne) We have had good results with EPISODE on problems with a solution that looks like a square wave.

(Hindmarsh) It could be that the solution is in a 'semi-stable' state of equilibrium and that certain perturbations can make it go unstable. There is an example of such a semi-stable problem in our paper presented at this meeting [and in this volume], and the difficulty was easily controlled by tightening the error control appropriately.

QUESTION: Dr. Hindmarsh, have you considered merging the several variants of GEAR into a single code?

ANSWER: (Hindmarsh) We have thought about it, but it's not an easy task and the desirability of it is not clear either. We would have to decide how many variants to include (many are conceivable), and whether to burden the user with all of the modules he doesn't need at any given time.

QUESTION: Would you comment on the progress of proving the correctness of a code? For example, can one show that every statement is executable?

ANSWERS: (Krogh) There is software available at JPL and at other places which counts the number of times each statement in a code is executed.

(Shampine) We also have software at Sandia which counts the number of times each statement in a code is executed as a problem is solved. Attempting to exercise every executable statement proved valuable when writing STEP. For example, we found a statement which could not be executed.

QUESTION: Aren't correctness and efficiency (or timing) of code both machine dependent?

ANSWER: (Hull) Yes, but in testing one would strive to determine relative merits of programs. To a certain extent, the relative ordering of programs would be machine independent.

Subject Index